T0137036

Studies in Computational Intelligence

Volume 729

Series editor

Janusz Kacprzyk, Polish Academy of Sciences, Warsaw, Poland
e-mail: kacprzyk@ibspan.waw.pl

About this Series

The series "Studies in Computational Intelligence" (SCI) publishes new developments and advances in the various areas of computational intelligence—quickly and with a high quality. The intent is to cover the theory, applications, and design methods of computational intelligence, as embedded in the fields of engineering, computer science, physics and life sciences, as well as the methodologies behind them. The series contains monographs, lecture notes and edited volumes in computational intelligence spanning the areas of neural networks, connectionist systems, genetic algorithms, evolutionary computation, artificial intelligence, cellular automata, self-organizing systems, soft computing, fuzzy systems, and hybrid intelligent systems. Of particular value to both the contributors and the readership are the short publication timeframe and the worldwide distribution, which enable both wide and rapid dissemination of research output.

More information about this series at http://www.springer.com/series/7092

Yaniv Altshuler · Alex Pentland
Alfred M. Bruckstein

Swarms and Network Intelligence in Search

 Springer

Yaniv Altshuler
MIT Media Lab
Massachusetts Institute of Technology
Cambridge, MA
USA

Alfred M. Bruckstein
Computer Science Department
Israeli Institute of Technology (Technion)
Haifa
Israel

Alex Pentland
MIT Media Lab
Massachusetts Institute of Technology
Cambridge, MA
USA

ISSN 1860-949X   ISSN 1860-9503   (electronic)
Studies in Computational Intelligence
ISBN 978-3-319-87591-0   ISBN 978-3-319-63604-7   (eBook)
DOI 10.1007/978-3-319-63604-7

Printed on acid-free paper

This Springer imprint is published by Springer Nature
The registered company is Springer International Publishing AG
The registered company address is: Gewerbestrasse 11, 6330 Cham, Switzerland

Preface

Last decade has seen a paradigmatic change in the operational processes of modern armies' aerial forces, as designers and commanders have been gradually shifting their focus towards unmanned aerial vehicles (UAVs). Indeed, over 50% of the planes in the USA today are unmanned, with this trend expected to further increase, and be adopted by additional Western armies in the coming years. This has been the case in visual intelligent (VISINT), and recently also in tactical signal intelligence (SIGINT), which is traditionally in charge of 80% of the information gathered by the intelligence corps.

Therefore, as the use of Drones as an integral component on ongoing intelligence gathering in wartime, as well as during the "battle within the wars" increases, so grows the importance of the need to base this use on an efficient infrastructure. In other words, an innovative small-scale dedicated UAV squadron (namely, a Drones Swarm), designed for special missions, may function perfectly with high redundancy and inefficient use of its resources, but a regular large-scale information gathering that is based on unmanned vehicles operating in swarms cannot. Furthermore, the lack of an efficient infrastructure that assumes control of the low-level resource utilization tasks means that these tasks must ultimately be taken care of by the human operators (as is being done today)—dramatically reducing the number of tasks these can engage, increasing the time it takes them to do so, as well as the overall cost of this process, and ultimately significantly limiting the vehicles' operational potential.

As the complexity of the problem increases, so does the impact of optimizing capabilities on the overall resources required in order to guarantee a pre-defined level of performance. In other words, a successful use of large-scale swarms of UAVs as a combat and intelligence gathering tool necessitates the development of an efficient mechanism for optimization of their utilization, specifically in the design and maintenance of their patrolling routes.

This book offers a comprehensive analysis of the theory and tools needed for the development of an efficient and robust infrastructure for the design of collaborative patrolling UAV swarms, focusing on its applications for tactic intelligence Drones. The systems under discussion enable flocks of semi-autonomous vehicles to perform an ongoing dynamic efficient patrolling and scanning of pre-defined "search region", in a robust and near-optimal way, in as fast time as possible, while guaranteeing detection of all targets that are located in that region. Theoretical limitations of such systems are discussed, as well as the trade-offs between the various economic and operational parameters of the system.

Cambridge, USA   Yaniv Altshuler
Cambridge, USA   Alex Pentland
Haifa, Israel   Alfred M. Bruckstein

Contents

Introduction to Swarm Search

1 Overview

Last decade has seen a paradigmatic change in the operational processes of modern armies aerial forces, as designers and commanders have been gradually shifting their focus towards unmanned aerial vehicles (UAVs) [41, 41, 64, 126]. Indeed, over 50% of the planes in the US today are unmanned, with this trend expected to further increase, and be adopted by additional western armies in the coming years [45, 108]. This trend is not exclusive to superpowers such as the United States, as for example Israel, a country that is mainly focused on relatively limited conflicts and geographic arenas has become a leader in drones technology and their military adoption [68, 69, 85], and is currently relying heavily on the use of UAVs in its military force [104]. Such vehicles are becoming an increasingly dominant element of its intelligence platforms. This has been the case in visual intelligent (VISINT) [95], and recently also in tactical signal intelligence (SIGINT), which is traditionally in charge of 80% of the information gathered by the intelligence corps [5, 28]. Another frequent use-case of self-organizing drones swarm is for providing (and then, maintaining) continuous network connectivity availability both for military as well as civilian environments [87].

Therefore, as the use of Drones as an integral component on ongoing intelligence gathering in wartime, as well as during the "battle within the wars" increases, so grows the importance of the need to base this use on an efficient infrastructure [40]. In other words, an innovative small scale dedicated UAV squadron (namely, a Drones Swarm), designed for special missions may function perfectly with high redundancy and inefficient use of its resources, but a regular large-scale information gathering that is based on unmanned vehicles operating in swarms, cannot. Furthermore, the lack of an efficient infrastructure that assumes control of the low-level resource utilization tasks means that these tasks must ultimately be taken care of by the human operators (as is being done today) dramatically reducing the number of tasks these can engage, increasing the time it takes them to do so, as well as the overall cost of this process, and ultimately significantly limiting the vehicles operational potential.

© Springer International Publishing AG 2018
Y. Altshuler et al., *Swarms and Network Intelligence in Search*,
Studies in Computational Intelligence 729, DOI 10.1007/978-3-319-63604-7_1

In its military aspects, it is interesting to mention the roots of this field, dating back to *World War II* [84, 115]. The first planar search problem considered is the patrol of a corridor between parallel borders separated by a constant width. This problem aimed for determining optimal patrol strategies for aircraft searching for ships in a channel [72], with a more generic theory of optimal scanning that was later proposed in [93]. Another modern use of drones swarms for putting a blockade around cities is extensively discussed in [60].

As the complexity of the problem increases, so does the impact of optimizing capabilities on the overall resources required in order to guarantee a pre-defined level of performance. In other words, a successful use of large scale swarms of UAVs as a combat and intelligence gathering tool necessitates the development of an efficient mechanism for optimization of their utilization, specifically in the design and maintenance of their patrolling routes.

To facilitate the transition of such methods from theory to practical mechanisms designed and deployed in existing systems, we need to create a common language for use between researchers and practitioners in this new area, ranging from the theory of computational social sciences to conventional computer science and network engineering. This is also becoming increasingly important with the increase use of droned by terrorists [39, 114] which necessitates the continuous pursuit after new scientific edges, in order to preserve one's comparative technological advantage.

This book offers a comprehensive analysis of the theory and tools needed for the development of an efficient and robust infrastructure for the design of collaborative patrolling UAV swarms, focusing on its applications for tactic intelligence drones. The systems under discussion enable flocks of semi-autonomous vehicles to perform an ongoing dynamic efficient patrolling and scanning of pre-defined "search region", in a robust and near-optimal way, in as fast time as possible, while guaranteeing detection of all targets that are located in that region. Theoretical limitations of such systems are discussed, as well as the trade-offs between the various economic and operational parameters of the system.

2   Swarm Intelligence

There is a growing demand for robotics solutions to increasingly complex and varied challenges. In the course of time it was realized that often a single robot is not the best solution for many application domains. Instead, teams of robots are called upon to work in an intelligently coordinated fashion for achieving efficiency and reliability via redundancy.

In [50] a detailed description of swarm-robotics application domains is presented, demonstrating how large-scale decentralized systems of autonomous robotic agents can be significantly more effective than a single robot in many areas. However, when designing such systems it should be noticed that simply increasing the number of

robots assigned to a task does not necessarily improve the system's performance —
multiple robots must intelligently cooperate to avoid disturbing each other's activity
and achieve efficiency.

In nature, "simple minded" animals such as ants, bees or birds cooperate to achieve
common goals and exhibit amazing feats of collaborative work. It seems that these
animals are "programmed" to interact locally in such a way that the desired global
behavior is likely to emerge even if some individuals of the colony die or fail to
carry out their task for other reasons. A similar approach may be considered for
coordinating a group of robots without a central supervisor, by using only local
interactions between the robots. When this decentralized approach is used, much of
the communication overhead (typical of centralized systems) is saved, the hardware
of the robots can be fairly simple, and better modularity is achieved. A properly
designed system should be readily scalable, achieving reliability through redundancy.

There are several key advantages to the use of such *intelligent swarm robotics*.
First, such systems inherently enjoy the benefit of parallelism. In task-decomposable
application domains, robot teams can accomplish a given task more quickly than a
single robot, by dividing the task into sub-tasks and executing them concurrently. In
certain cases, a single robot may simply be unable to accomplish the task on its own
(e.g. to carry a large and heavy object).

Second, decentralized systems tend to be, by their very nature, much more robust
than centralized systems (or systems comprised of a single but very complex unit).
Generally speaking, a team of robots may provide a more robust solution by introduc-
ing redundancy, and by eliminating any single point of failure. While considering
the alternative of using a single sophisticated robot, we should note that even the
most complex and reliable robot may suffer an unexpected malfunction, which will
prevent it from completing its task. When using a multi agents system, on the other
hand, even if a large number of the agents stop working for some reason, the entire
group will often still be able to complete its task, although perhaps slower. For example, for exploring a hazardous region (such as a minefield or the surface of Mars),
the benefit of redundancy and robustness offered by a multi agents system is quite
obvious, and it is in this context that Rodney Brooks wrote his famous "Fast, cheap
and out of control" report [36].

Another advantage of the decentralized swarm approach is the ability of dynamically reallocating sub-tasks between the swarm's units, thus adapting to unexpected
changes in the environment. Furthermore, since the system is decentralized, it can
respond relatively quickly to such changes, due to the benefit of locality, meaning —
the ability to swiftly respond to changes without the need of notifying a hierarchical
"chain of command". Note that as the swarm becomes larger, this advantage becomes
increasingly important.

In addition to the ability of quick response to changes, the decentralized nature
of such systems also improves their scalability. The scalability of multi-agent systems is derived from relying on the "emergence" of task completion by inherently
low communication and computation overhead protocol implemented by the agents.
As the tasks assigned nowadays to multi-agent based systems become increasingly
complex, so does the importance of the high scalability of the systems.

Finally, by using heterogeneous swarms, even more efficient systems could be designed, thanks to the utilization of different types of agents whose physical properties enable them to perform much more efficiently in certain special tasks.

Significant research effort has been invested during the last few years in design and simulation of multi-agent robotics and intelligent swarm systems (see e.g. [34, 44, 48, 65, 66, 80, 120, 121]).

Such designs are often inspired by biology (see [54, 71] for evolutionary algorithms, [26] or [35, 129] for behavior based control models, [49, 51, 81, 101, 112] for flocking and dispersing models, [29, 63, 124] for predator-prey approaches), by physics [46, 61, 70], sociology [73, 106, 116], network theory [18, 23, 24, 97] or by economics applications [6, 20–22, 57, 76, 83, 90, 105, 125, 132].

A swarm based robotics system can generally be defined as a highly decentralized group of extremely simple robotic agents, with limited communication, computation and sensing abilities, designed and deployed to accomplish various tasks. Tasks that have been of particular interest to researchers in recent years include synergetic mission planning [7, 117], patrolling [2, 3, 17], fault tolerance cooperation [74, 91, 127], network security [99], crowds modeling [19], swarm control [47, 82], human design of mission plans [78, 79], role assignment [42, 43, 89, 111, 133], multi-robot path planning [3, 55, 77, 107, 113, 128], traffic control [4, 96], formation generation [25, 32, 56, 59], formation keeping [27, 30, 123], exploration and mapping [94, 100, 105], target tracking [62, 92], collaborative cleaning [10, 13, 16, 122], control architecture for drones swarm [11, 15] and target search [67, 75].

Unfortunately, the mathematical\geometrical theory of such multi-agents systems is far from being satisfactory, as pointed out in [31, 33, 53, 86] and many other papers.

Our interest is focused on developing mathematical tools necessary for the design and analysis of such systems. For example, in [118] it was shown that a number of agents can arrange themselves equidistantly in a row via a sequence of linear adjustments, based on a simple "local" interaction. The convergence of the configuration to the desired one is exponentially fast. A different way of cooperation between agents, inspired by the behavior of ant-colonies, is described in [37]. There it was proved that a sequence of ants engaged in deterministic chain pursuit will find the shortest (i.e. straight) path from the ant-hill to the food source, using only local interactions. In [38] the behavior of a group of agents on $\mathbf{Z}^2$ is investigated, where each ant-like agent is pursuing his predecessor, according to a discrete biased-random-walk model of pursuit on the integer grid. The average paths of such a sequence of a(ge)nts engaged in a chain of probabilistic pursuit was shown to converge to the "straight line" between the origin and destination, and this too happens exponentially fast.

An in-depth analysis of the effect of certain geometric properties on the search efficiency of a collaborative swarm of autonomous drones appears in [12, 14], whereas an example of an analytic complexity bounds for this problem can be found in [8, 9]. A work that analyzed the effect of a stochastic framework for the same problem is presented in [98].

3 Building Intelligent Swarms Using Stupid Drones

A key principle in the notion of swarms, or multi agent robotics, is the simplicity of the individual drones. The notion of "simplicity" here means that the drones should be *significantly simpler* than a "single sophisticated robotic system", which can be constructed for the same purpose. As a result, the capabilities and the resources of such simple agents are assumed to be very limited, with respect to the following aspects:

- Memory resources — basic agents should be assumed to contain only $O(1)$ memory resources (i.e. the size of memory is independent of the size of the problem or the number of agents). This usually imposes many interesting limitations on the agents. For example, agents can remember only a limited history of their activities so far. Thus, protocols designed for agents with such limited memory resources are usually very simple and attempt to solve a given problem by relying on some (necessarily local) basic patterns arising in the environment. The task is completed by a repetition of these patterns by a large number of agents.
- Sensing capabilities — defined according to the specific nature of the problem. For example, for agents moving along a 100×100 grid, a reasonable sensing radius may be 3 or 4, but certainly not 40.
- Computational resources — although agents are assumed to employ only limited computational resources, a formal definition of this constraint is hard to define. In general, most of the time-polynomial algorithms may be used, provided that the amount of memory the agents have is sufficient.
- Communication is very limited too — the issue of communication in multi agents systems has been extensively studied in recent years. Distinctions between implicit and explicit communication are usually made, in which implicit communication occurs as a side effect of other actions, or "through the world" (see, for example [88]), whereas explicit communication is a specific act intended solely to convey information to other robots on the team. Explicit communication can be performed in several ways, such as a short range point to point communication, a global broadcast, or by using some sort of distributed shared memory. Such memory is often referred to as a *pheromone*, used to convey small amounts of information between the agents [54, 119, 130, 131]. This approach is inspired from the coordination and communication methods used by many social insects — studies on ants (e.g. [1, 58]) show that the pheromone based search strategies used by ants in foraging for food in unknown terrains tend to be very efficient. Additional information can be found in the relevant NASA survey, focusing on "intelligent swarms" comprised of multiple "stupid satellites" [102, 103] or the following survey conducted by the US Naval Research Center [109].

In the spirit of designing a system which uses as simple agents as possible, we aspire that the agents will have as little communication capabilities as possible. With respect to the taxonomy of multi-agents discussed in [52], we would be interested in using agents of the types *COM-NONE* or if necessary of type *COM-NEAR* with respect to their communication distances, and of types *BAND-MOTION*,

BAND-LOW or even *BAND-NONE* (if possible) with respect to their communication bandwidth. Therefore, although a certain amount of implicit communication can hardly be avoided (due to the simple fact that by changing the environment, the agents are constantly generating some kind of implicit information), explicit communication should be strongly limited or avoided altogether, in order to fit our paradigm (note that in many works in this field, this is not the case, and communication, as well as memory, resources, are often being used in order to create complex cooperative systems).

In summary, while designing intelligent swarm systems we must assume (and often even *aspire for*) having an available individual drones that are myopic, mute, senile and rather stupid.

4 Organization

The rest of this book is divided into three parts covering three complementary themes and is structured as follows. The first part contains four studies that touch on fundamental aspects of *swarm search* in the grid domain, raising and discussing topics such as the conceptual definition of the problem, the drones' capabilities, and the interplay between the two. The second part of the book quickly touches the topic of stochasticity in swarm systems, analyzing as a case study the same problem that is analyzed in the previous part, under stochastic regime. The third section focuses directly on the use of a decentralized swarm of Unmanned Aerial Vehicles (or UAVs) for the purpose of scanning a pre-defined area for a group of evading targets, in the most efficient way possible. In this part two approaches are discussed, one that relies on the geometry of the problem, and another that takes a network-oriented approach.

Chapter 2 analyzes the behavior of a swarm of autonomous robotic agents, or drones, designed for cooperatively explore an unknown area (for purposes of cleaning, painting, etc.). Each robot is assumed to acquire only the information which is available in its immediate vicinity, and as the only way of inter-robot communication is by leaving traces on the common ground and sensing the traces left by other robots. This subsequently means that a successful completion of the exploration can only be guaranteed through an efficient collaboration at the swarm level, that emerge from the multitude of local interactions. A protocol designed for searching a pre-defined area that guarantees task completion (unless all robots die) is presented, and an upper bound on the time complexity of this protocol is given. In addition, extensive simulations of the problem are presented on several types of search–regions. These simulations indicate that the precise completion time depends on the number of robots, their initial locations, and the shape of the region to be explored.

Chapter 3 discusses the dynamic generalization of the exploration problem that is presented in Chap. 2, involving a swarm of collaborative drones that are required to search a dynamically expanding region on the Z^2 grid (whose "pixels" that were visited by the drones can become "un-searched" after a certain period of time). The goal of the swarm members is to "clean" the spreading contamination in as little time as possible. In this work we present an algorithm-agnostic lower bound for the problem, as well as a collaborative swarm search algorithm, accompanied with a variety of experimental results.

Chapter 4 continues the discussion about this dynamic collaborative search problem, presenting a collaborative swarm search algorithm, as well as several upper bounds on the completion time it takes a swarm of k drones, of various velocities, to execute it until the search throughout the dynamic search area is completed. These upper bounds are also visually demonstrated using several examples of search-regions.

Chapter 5 proceeds to examine the complexity of the collaborative search problem in dynamic environments, by analyzing the upper and lower bounds that were demonstrated in Chaps. 2–4. Although these bounds yield an (analytically provable) exact numeric bounds (albeit, certainly not tight ones) transforming those into complexity expressions provides an additional value, as it enables to discuss the "difficulty" of the problem, regardless of its specific characterization, as well as state mandatory requirements from any swarm of search drones, necessary for the successful completion of the search mission (and again, regardless of the exact numeric specifications of the problem). Specifically, it is shown that regardless of the algorithm used, and the drones' hardware and software specifications, the minimal number of drones required in order for such coverage to be possible is $\Omega(\sqrt{n})$ (n being the size of the search region). In addition, it is shown that when the region expands at a sufficiently slow rate, a team of $\Theta(\sqrt{n})$ robots can cover it in at most $O(n^2 \ln n)$ time. Furthermore, this completion time can even be achieved by myopic robots, with no ability to directly communicate with each other, and where each robot is equipped with a memory of size $O(1)$ bits with respect to the size of the region (namely, when the robots cannot maintain maps of the terrain, nor plan complete paths). Regarding the coverage of non-expanding regions in the grid, the current best known result of $O(n^2)$ is improved by demonstrating an algorithm that guarantees such a coverage with completion time of $O(\frac{1}{k}n^{1.5} + n)$ in the worst case, and faster for shapes of perimeter length which is shorter than $O(n)$.

Chapter 6 introduces a stochastic regime to the search problem, assuming that the shape and size of the search area is unknown in advance, and expands outward with probability p at every time step. Probabilistic lower bounds for the minimal number of drones required in order to guarantee a successful completion of the search mission is shown, as well a bound over the minimal time required to enable a collaborative coverage of the expanding region, regardless of the algorithm used and the drones hardware and software specifications. In addition, a method of producing ad-hoc lower bounds, for any given desired correctness probability is presented. This chapter further presents impossibility results that for any given number of drones available, and spreading probability, provides an upper bound for the minimal value of the initial area of the expanding region which is guaranteed to be impossible to clear. These analytic bounds are also supported empirical computer simulation results.

In Chap. 7 the *Cooperative Hunters* problem is presented, where a swarm of Unmanned Air Vehicles (UAVs) is used for searching after one or more "evading targets", which freely maneuver in a predefined area while trying to avoid detection by the swarm drones. By arranging themselves into an efficient geometric collaborative flight formation, the drones optimize their integrated sensing capabilities, enabling the completion of a successful search of a rectangular territory. This designed is shown to be able to guarantee the detection of the targets, even in cases where the targets are faster than the swarm drones and have better sensors. This is achieved through the inherent scalability of the proposed design which can compensate any addition to the targets ability to maneuver or foresee the behavior of the drones with an increase in the number of drones.

Chapter 8 concludes the book, by presenting a network-oriented approach for the problem of using a swarm of search agents in order to maximize the detection probability of one or more targets, who freely move through a pre-defined network (e.g. roads, internet links, etc.). This is demonstrated as the optimization of large scale swarms of reconnaissance drones capable of producing on-demand optimal coverage strategies for any given search scenario. Given an estimation cost of the threat potential damages, as well as types of monitoring drones available and their comparative performance, the proposed method generates an analytically provable strategy, stating the optimal number and types of drones to be deployed, in order to costefficiently monitor a pre-defined region for targets maneuvering using a given roads networks. The proposed model is demonstrated using a unique dataset of the Israeli transportation network, on which different deployment schemes for drones deployment are evaluated.

5 Conclusion

In summary, we would like to cite a statement made by a scientist after watching an ant making his laborious way across a wind-and-wave-molded beach [110]:

> An ant, viewed as a behaving system, is quite simple. The apparent complexity of its behavior over time is largely a reflection of the environment in which it finds itself.

Such a point of view, as well as the results of our analysis and simulations, lead us to believe that even simple, ant-like drones, when allowed to synergically collaborate, can yield a complicated, adaptive and quite efficient macroscopic behavior, in the intelligent swarm-level scope.

References

1. F.R. Adler, D.M. Gordon, Information collection and spread by networks of partolling agents. Am. Nat. **140**(3), 373–400 (1992)
2. N. Agmon, S. Kraus, G.A. Kaminka, Multi-robot perimeter patrol in adversarial settings, in *IEEE International Conference on Robotics and Automation (ICRA 2008)* (2008), pp. 2339–2345
3. N. Agmon, V. Sadov, G.A. Kaminka, S. Kraus, The impact of adversarial knowledge on adversarial planning in perimeter patrol, in *AAMAS '08: Proceedings of the 7th international*

joint conference on Autonomous agents and multiagent systems (International Foundation for Autonomous Agents and Multiagent Systems, Richland, SC, 2008), pp. 55–62

4. A. Agogino, K. Tumer, Regulating air traffic flow with coupled agents, in *AAMAS '08: Proceedings of the 7th international joint conference on Autonomous agents and multiagent systems* (International Foundation for Autonomous Agents and Multiagent Systems, Richland, SC, 2008), pp. 535–542

5. M.M. Aid, All glory is fleeting: sigint and the fight against international terrorism. Intell. Natl. Secur. **18**(4), 72–120 (2003)

6. S. Aknine, O. Shehory, A feasible and practical coalition formation mechanism leveraging compromise and task relationships, in *Proceedings of the IEEE/WIC/ACM international conference on Intelligent Agent Technology* (2006), pp. 436–439

7. R. Alami, S. Fleury, M. Herrb, F. Ingrand, F. Robert, Multi-robot cooperation in the martha project. IEEE Robot. Autom. Mag. **5**(1), 36–47 (1998)

8. Y. Altshuler, A.M. Bruckstein, The complexity of grid coverage by swarm robotics, in *ANTS 2010* (LNCS, 2010), pp. 536–543

9. Y. Altshuler, A.M. Bruckstein, Static and expanding grid coverage with ant robots: complexity results. Theoret. Comput. Sci. **412**(35), 4661–4674 (2011)

10. Y. Altshuler, A.M. Bruckstein, I.A. Wagner, Swarm robotics for a dynamic cleaning problem, in *IEEE Swarm Intelligence Symposium* (2005), pp. 209–216

11. Y. Altshuler, V. Yanovski, I.A. Wagner, A.M. Bruckstein, The cooperative hunters - efficient cooperative search for smart targets using uav swarms, in *Second International Conference on Informatics in Control, Automation and Robotics (ICINCO), the First International Workshop on Multi-Agent Robotic Systems (MARS)* (2005), pp. 165–170

12. Y. Altshuler, I.A. Wagner, A.M. Bruckstein, Shape factor's effect on a dynamic cleaners swarm, in *Third International Conference on Informatics in Control, Automation and Robotics (ICINCO), the Second International Workshop on Multi-Agent Robotic Systems (MARS)* (2006), pp. 13–21

13. Y. Altshuler, V. Yanovsky, I. Wagner, A. Bruckstein, Swarm intelligence searchers, cleaners and hunters. *Swarm Intelligent Systems* (2006), pp. 93–132

14. Y. Altshuler, I.A. Wagner, A.M. Bruckstein, On swarm optimality in dynamic and symmetric environments **7**, 11 (2008)

15. Y. Altshuler, V. Yanovsky, A.M. Bruckstein, I.A. Wagner, Efficient cooperative search of smart targets using uav swarms. ROBOTICA **26**, 551–557 (2008)

16. Y. Altshuler, I.A. Wagner, A.M. Bruckstein, Collaborative exploration in grid domains, in *Sixth International Conference on Informatics in Control, Automation and Robotics (ICINCO)* (2009)

17. Y. Altshuler, I.A. Wagner, V. Yanovski, A.M. Bruckstein, Multi-agent cooperative cleaning of expanding domains. Int. J. Robot. Res. **30**, 1037–1071 (2010)

18. Y. Altshuler, S. Dolev, Y. Elovici, N. Aharony, Ttled random walks for collaborative monitoring, in *NetSciCom 2010 (Second International Workshop on Network Science for Communication Networks), San Diego, CA, USA*, vol. 3 (2010)

19. Y. Altshuler, M. Fire, N. Aharony, Y. Elovici, A Pentland, How many makes a crowd? on the correlation between groups' size and the accuracy of modeling, in *International Conference on Social Computing, Behavioral-Cultural Modeling and Prediction* (Springer, 2012), pp. 43–52

20. Y. Altshuler, E. Shmueli, G. Zyskind, O. Lederman, N. Oliver, A. Pentland, Campaign optimization through behavioral modeling and mobile network analysis. IEEE Trans. Comput. Soc. Syst. **1**(2), 121–134 (2014)

21. Y. Altshuler, A. Sandy Pentland, G. Gordon, Social behavior bias and knowledge management optimization, in *Social Computing, Behavioral-Cultural Modeling, and Prediction* (Springer, 2015), pp. 258–263

22. Y. Altshuler, E. Shmueli, G. Zyskind, O. Lederman, N. Oliver, A.S. Pentland, Campaign optimization through mobility network analysis. *Geo-Intelligence and Visualization Through Big Data Trends* (2015), pp. 33–74

23. Y. Altshuler, A. Pentland, S. Bekhor, Y. Shiftan, A. Bruckstein, Optimal dynamic coverage infrastructure for large-scale fleets of reconnaissance uavs (2016), arXiv:1611.05735

24. Y. Altshuler, R. Puzis, Y. Elovici, S. Bekhor, A. Sandy Pentland, On the rationality and optimality of transportation networks defense: a network centrality approach. *Securing Transportation Systems* (2015), pp. 35–63

25. T. Arai, H. Ogata, T. Suzuki, Collision avoidance among multiple robots using virtual impedance, in *Proceedings of the IEEE/RSJ International Conference on Intelligent Robots and Systems* (1989), pp. 479–485

26. R.C. Arkin, T. Balch, Aura: principles and practice in review. J. Exp. Theor. Artif. Intell. **9**(2/3), 175–188 (1997)

27. T. Balch, R. Arkin, Behavior-based formation control for multi-robot teams. IEEE Trans. Robot. Autom. **14**(6), 926–939 (1998)

28. D. Ball et al., *Burma's Military Secrets: Signals Intelligence (SIGINT) from 1941 to Cyber Warfare* (White Lotus Press, 1998)

29. M. Benda, V. Jagannathan, R. Dodhiawalla, On optimal cooperation of knowledge sources. *Technical Report BCS-G2010-28, Boeing AI Center* (1985)

30. K. Bendjilali, F. Belkhouche, B. Belkhouche, Robot formation modelling and control based on the relative kinematics equations. Int. J. Robot. Autom. **24**(1), 79–88 (2009)

31. G. Beni, J. Wang, Theoretical problems for the realization of distributed robotic systems, in *IEEE Internal Conference on Robotics and Automation* (1991), pp. 1914–1920

32. R.M. Bhatt, C.P. Tang, V.N. Krovi, Formation optimization for a fleet of wheeled mobile robots a geometric approach. Robot. Auton. Syst. **57**(1), 102–120 (2009)

33. E. Bonabeau, M. Dorigo, G. Theraulaz, *Swarm Intelligence: From Natural to Artificial Systems* (Oxford University Press, US, 1999)

34. M. Brambilla, E. Ferrante, M. Birattari, M. Dorigo, Swarm robotics: a review from the swarm engineering perspective. Swarm Intell. **7**(1), 1–41 (2013)

35. R.A. Brooks, A robust layered control system for a mobile robot. IEEE J. Robot. Autom. **RA–2**(1), 14–23 (1986)

36. R.A. Brooks, A.M. Flynn, Fast, cheap and out of control, a robot invasion of the solar system. J. Br. Interplanet. Soc. **42**, 478–485 (1989)

37. A.M. Bruckstein, Why the ant trails look so straight and nice. Math. Intell. **15**(2), 59–62 (1993)

38. A.M. Bruckstein, C.L. Mallows, I.A. Wagner, Probabilistic pursuits on the integer grid. Am. Math. Mon. **104**(4), 323–343 (1997)

39. R.J. Bunker, Terrorist and insurgent unmanned aerial vehicles: Use, potentials, and military implications. Technical report, DTIC Document (2015)

40. D.J. Bursac, *Autonomous robotic weapons: Us army innovation for ground combat in the twenty-first century* (Technical report, US Army School for Advanced Military Studies Fort Leavenworth United States, 2015)

41. D. Byman, Why drones work. Foreign Aff. **92**(4), 32–43 (2013)

42. C. Candea, H. Hu, L. Iocchi, D. Nardi, M. Piaggio, Coordinating in multi-agent robocup teams. Robot. Auton. Syst. **36**(2–3), 67–86 (2001)

43. G. Chalkiadakis, C. Boutilier, Sequential decision making in repeated coalition formation under uncertainty, in *Proceedings of the 7th international joint conference on Autonomous agents and multiagent systems* (2008), pp. 347–354

44. G. Chalkiadakis, E. Markakis, C. Boutilier, Coalition formation under uncertainty: bargaining equilibria and the bayesian core stability concept, in *AAMAS '07: Proceedings of the 6th International Joint Conference on Autonomous Agents and Multiagent Systems* (ACM, New York, 2007), pp. 1–8

45. R. Chesney, Military-intelligence convergence and the law of the title 10/title 50 debate. J. Nat. Secur. Law Policy **5**, 539 (2012)

46. D. Chevallier, S. Payandeh, On kinematic geometry of multi-agent manipulating system based on the contact force information, in *The Sixth International Conference on Intelligent Autonomous Systems (IAS-6)* (2000), pp. 188–195

47. R. Connaughton, P. Schermerhorn, M. Scheutz, Physical parameter optimization in swarms of ultra-low complexity agents, in *AAMAS '08: Proceedings of the 7th International Joint Conference on Autonomous Agents and Multiagent Systems* (International Foundation for Autonomous Agents and Multiagent Systems, Richland, SC, 2008), pp. 1631–1634
48. S.A. DeLoach, M. Kumar, Multi-agent systems engineering: an overview and case study, *Intelligence Integration in Distributed Knowledge Management* (Idea Group Inc (IGI), 2008), pp. 207–224
49. J. Deneubourg, S. Goss, G. Sandini, F. Ferrari, P. Dario, Self-organizing collection and transport of objects in unpredictable environments, in *Japan-U.S.A. Symposium on Flexible Automation* (1990), pp. 1093–1098
50. M.B. Dias, A. Stentz, A market approach to multirobot coordination. *Technical Report, CMU-RI - TR-01-26, Robotics Institute, Carnegie Mellon University* (2001)
51. A. Drogoul, J. Ferber, From tom thumb to the dockers: some experiments with foraging robots, in *Proceedings of the Second International Conference on Simulation of Adaptive Behavior* (1992), pp. 451–459
52. G. Dudek, M.R.M. Jenkin, E. Milios, D. Wilkes, A taxonomy for multi-agent robotics. Auton. Robots J. **3**(4), 375–397 (1996)
53. A. Efraim, D. Peleg, Distributed algorithms for partitioning a swarm of autonomous mobile robots, *Structural Information and Communication Complexity*, vol. 4474, Lecture Notes in Computer Science (2007), pp. 180–194
54. A. Felner, Y. Shoshani, Y. Altshuler, A.M. Bruckstein, Multi-agent physical a* with large pheromones. J. Auton. Agents Multi-Agent Syst. **12**(1), 3–34 (2006)
55. C. Ferrari, E. Pagello, J. Ota, T. Arai, Multirobot motion coordination in space and time. Robot. Auton. Syst. **25**, 219–229 (1998)
56. J. Fredslund, M.J. Mataric, Robot formations using only local sensing and control, in *Proceedings of the International Symposium on Computational Intelligence in Robotics and Automation (IEEE CIRA 2001)* (2001), pp. 308–313
57. B.P. Gerkey, M.J. Mataric, *Sold! market methods for multi-robot control* (IEEE Trans. Robot. Autom. Spec, Issue Multi-robot Syst, 2002)
58. D.M. Gordon, The expandable network of ant exploration. Anim. Behav. **50**, 372–378 (1995)
59. N. Gordon, I.A. Wagner, A.M. Bruckstein, Discrete bee dance algorithms for pattern formation on a grid, in *IEEE International Conference on Intelligent Agent Technology (IAT03)* (2003), pp. 545–549
60. S. Graham, *Cities under siege: The new military urbanism* (Verso Books, 2011)
61. J. Hagelbäck, S.J. Johansson, Demonstration of multi-agent potential fields in real-time strategy games, in *AAMAS '08: Proceedings of the 7th International Joint Conference on Autonomous Agents and Multiagent Systems* (International Foundation for Autonomous Agents and Multiagent Systems, Richland, SC, 2008), pp. 1687–1688
62. I. Harmatia, K. Skrzypczykb, Robot team coordination for target tracking using fuzzy logic controller in game theoretic framework. Robot. Auton. Syst. **57**(1), 75–86 (2009)
63. T. Haynes, S. Sen, Adaptation and learning in multi-agent systems. *Evolving Behavioral Strategies in Predators and Prey*, vol. 1042, Lecture Notes in Computer Science (Springer, Berlin, 1986), pp. 113–126
64. I. Henderson, Civilian intelligence agencies and the use of armed drones, in *Yearbook of International Humanitarian Law-2010* (Springer, 2011), pp. 133–173
65. S. Hettiarachchi, W. Spears, Moving swarm formations through obstacle fields, in *International Conference on Artificial Intelligence* (2005)
66. M.G. Hinchey, R. Sterritt, C. Rouff, Swarms and swarm intelligence. Computer **40**(4), 1 (2007)
67. G. Hollinger, S. Singh, J. Djugash, A. Kehagias, Efficient multi-robot search for a moving target. Int. J. Robot. Res. **28**(2), 201–219 (2009)
68. D.E. Johnson, *Military capabilities for hybrid war: Insights from the Israel Defense Forces in Lebanon and Gaza* (Rand Corporation, 2014)

69. Lt. Kendra L.B. Cook, The silent force multiplier: the history and role of uavs in warfare, in *Aerospace Conference, 2007 IEEE* (IEEE, 2007), pp. 1–7
70. S. Kirkpatrick, J.J. Schneider, How smart does an agent need to be? Int. J. Mod. Phys. C **16**, 139–155 (2005)
71. T. Klos, G.J. van Ahee, Evolutionary dynamics for designing multi-period auctions, in *AAMAS '08: Proceedings of the 7th international joint conference on Autonomous agents and multiagent systems* (International Foundation for Autonomous Agents and Multiagent Systems, Richland, SC, 2008), pp. 1589–1592
72. B. Koopman, *Search and Screening: General Principles with Historical Applications* (Pergamon Press, 1980)
73. P.M. Krafft, J. Zheng, W. Pan, N. Della Penna, Y. Altshuler, E. Shmueli, J.B. Tenenbaum, A. Pentland, Human collective intelligence as distributed bayesian inference (2016), arXiv:1608.01987
74. S. Kraus, O. Shehory, G. Taase, Coalition formation with uncertain heterogeneous information, in *Proceedings of the second international joint conference on Autonomous agents and multiagent systems* (2003), pp. 1–8
75. S.M. LaValle, D. Lin, L.J. Guibas, J.C. Latombe, R. Motwani, Finding an unpredictable target in a workspace with obstacles, in *Proceedings of the 1997 IEEE International Conference on Robotics and Automation (ICRA-97)* (1997), pp. 737–742
76. Y.-Y. Liu, J.C. Nacher, T. Ochiai, M. Martino, Y. Altshuler, Prospect theory for online financial trading. PLoS ONE **9**(10), e109458 (2014)
77. V.J. Lumelsky, K.R. Harinarayan, Decentralized motion planning for multiple mobile robots: the cocktail party model. Auton. Robots **4**(1), 121–136 (1997)
78. D. MacKenzie, R. Arkin, J. Cameron, Multiagent mission specification and execution. Auton. Robots **4**(1), 29–52 (1997)
79. E. Manisterski, R. Lin, S. Kraus, Understanding how people design trading agents over time, in *AAMAS '08: Proceedings of the 7th international joint conference on Autonomous agents and multiagent systems* (International Foundation for Autonomous Agents and Multiagent Systems, Richland, SC, 2008), pp. 1593–1596
80. S. Mastellone, D.M. Stipanovi, C.R. Graunke, K.A. Intlekofer, M.W. Spong, Formation control and collision avoidance for multi-agent non-holonomic systems: theory and experiments. Int. J. Robot. Res. **27**(1), 107–126 (2008)
81. M.J. Mataric, Designing emergent behaviors: From local interactions to collective intelligence, in *Proceedings of the Second International Conference on Simulation of Adaptive Behavior*, ed. by J. Meyer, H. Roitblat, S. Wilson (MIT Press, 1992), pp. 432–441
82. M.J. Mataric, *Interaction and Intelligent Behavior*. Ph.D. thesis, Massachusetts Institute of Technology (1994)
83. N. Michael, M.M. Zavlanos, V. Kumar, G.J. Pappas, Distributed multi-robot task assignment and formation control, in *IEEE International Conference on Robotics and Automation, 2008 (ICRA 2008)* (2008), pp. 128–133
84. P.M. Morse, G.E. Kimball, *Methods of Operations Research* (MIT Press, Wiley, New York, 1951)
85. F. Naaz, Indo-israel military cooperation. Strateg. Anal. **24**(5), 969–985 (2000)
86. R. Olfati-Saber, Flocking for multi-agent dynamic systems: algorithms and theory. IEEE Trans. Autom. Control **51**(3), 401–420 (2006)
87. D. Orfanus, E.P. de Freitas, F. Eliassen, Self-organization as a supporting paradigm for military uav relay networks. IEEE Commun. Lett. **20**(4), 804–807 (2016)
88. E. Pagello, A.D. Angelo, F. Montesello, F. Garelli, C. Ferrari, Cooperative behaviors in multi-robot systems through implicit communication. Robot. Auton. Syst. **29**(1), 65–77 (1999)
89. E. Pagello, A.D. Angelo, C. Ferrari, R. Polesel, R. Rosati, A. Speranzon, Emergent behaviors of a robot team performing cooperative tasks. *Advanced Robotics* (2002)
90. W. Pan, Y. Altshuler, A. Pentland, Decoding social influence and the wisdom of the crowd in financial trading network, in *Privacy, Security, Risk and Trust (PASSAT), 2012 International Conference on and 2012 International Confernece on Social Computing (SocialCom)* (IEEE, 2012), pp. 203–209

91. L.E. Parker, Alliance: an architecture for fault-tolerant multi-robot cooperation. IEEE Trans. Robot. Autom. **14**(2), 220–240 (1998)
92. L.E. Parker, C. Touzet, Multi-robot learning in a cooperative observation task. Distrib. Auton. Robot. Syst. **4**, 391–401 (2000)
93. K. Passino, M. Polycarpou, D. Jacques, M. Pachter, Y. Liu, Y. Yang, M. Flint, M. Baum, *Cooperative Control for Autonomous Air Vehicles, chapter Cooperative Control and Optimization* (Kluwer Academic, Boston, 2002)
94. M. Pfingsthorn, B. Slamet, A. Visser, A scalable hybrid multi-robot SLAM method for highly detailed maps. *RoboCup 2007: Robot Soccer World Cup XI*, vol. 5001 *Lecture Notes in Computer Science* (Springer, Berlin, 2008), pp. 457–464
95. J.T.K. Ping, A.E. Ling, T.J. Quan, C. Yea Dat, Generic unmanned aerial vehicle (uav) for civilian application-a feasibility assessment and market survey on civilian application for aerial imaging, in *2012 IEEE Conference on Sustainable Utilization and Development in Engineering and Technology (STUDENT)* (IEEE, 2012), pp. 289–294
96. S. Premvuti, S. Yuta, Consideration on the cooperation of multiple autonomous mobile robots, in *Proceedings of the IEEE International Workshop of Intelligent Robots and Systems* (1990), pp. 59–63
97. R. Puzis, Y. Altshuler, Y. Elovici, S. Bekhor, Y. Shiftan, A.S. Pentland, Augmented betweenness centrality for environmentally-aware traffic monitoring in transportation networks
98. E. Regev, Y. Altshuler, A.M. Bruckstein, The cooperative cleaners problem in stochastic dynamic environments (2012), arXiv:1201.6322
99. M. Rehak, M. Pechoucek, P. Celeda, V. Krmicek, M. Grill, K. Bartos, Multi-agent approach to network intrusion detection, in *AAMAS '08: Proceedings of the 7th international joint conference on Autonomous agents and multiagent systems* (International Foundation for Autonomous Agents and Multiagent Systems, Richland, SC, 2008), pp. 1695–1696
100. I.M. Rekleitis, G. Dudek, E. Milios, Experiments in free-space triangulation using cooperative localization, in *IEEE/RSJ/GI International Conference on Intelligent Robots and Systems (IROS)* (2003)
101. W. Ren, N. Sorensena, Distributed coordination architecture for multi-robot formation control. Robot. Auton. Syst. **56**(4), 324–333 (2008)
102. C. Rouff, W. Truszkowski, J. Rash, M. Hinchey, Formal approaches to intelligent swarms, in *Software Engineering Workshop, 2003. Proceedings. 28th Annual NASA Goddard* (IEEE, 2003), pp. 51–57
103. C.A. Rouff, W.F. Truszkowski, J.L. Rash, M.G. Hinchey, A survey of formal methods for intelligent swarms (NASA Goddard Space Flight Center, Greenbelt, MD, 2005)
104. R. Sanders, An israeli military innovation: Uavs. Technical report, DTIC Document (2003)
105. S. Sariel, T. Balch, Real time auction based allocation of tasks for multi-robot exploration problem in dynamic environments, in *Proceedings of the AAAI-05 Workshop on Integrating Planning into Scheduling* (2005), pp. 27–33
106. S. Savarimuthu, M. Purvis, M. Purvis, Tag based model for knowledge sharing in agent society, discussion paper 2009/01. Discussion paper, Department of Information Science, University of Otago, Dunedin, New Zealand (2009)
107. R. Sawhney, K.M. Krishna, K. Srinathan, M. Mohan, On reduced time fault tolerant paths for multiple uavs covering a hostile terrain, in *AAMAS '08: Proceedings of the 7th international joint conference on Autonomous agents and multiagent systems* (International Foundation for Autonomous Agents and Multiagent Systems, Richland, SC, 2008), pp. 1171–1174
108. N. Schörnig, Unmanned warfare: towards a neo-interventionist era? in *The Armed Forces: Towards a Post-Interventionist Era?* (Springer, 2013), pp. 221–235
109. A.C. Schultz, L.E. Parker, *Multi-robot Systems: From Swarms to Intelligent Automata: Proceedings from the 2002 NRL Workshop on Multi-robot Systems* (Springer Science & Business Media, 2013)
110. H.A. Simon, *The Sciences of the Artificial*, 2nd edn. (MIT Press, 1981)
111. P. Stone, M. Veloso, Task decomposition, dynamic role assignment, and low-bandwidth communication for real-time strategic teamwork. Artif. Intell. **110**(2), 241–273 (1999)

112. H. Su, X. Wang, Z. Lin, Flocking of multi-agents with a virtual leader. IEEE Trans. Autom. Control **54**(2), 293–307 (2009)
113. P. Svestka, M.H. Overmars, Coordinated path planning for multiple robots. Robot. Auton. Syst. **23**(3), 125–152 (1998)
114. M.T. Tedesco, Countering the unmanned aircraft systems threat. Mil. Rev. **95**(6), 64 (2015)
115. A. Thorndike, *Summary of antisubmarine warfare operations in world war ii* (Summary report, NDRC Summary Report, 1946)
116. G. Trajkovski, S.G. Collins, *Handbook of Research on Agent-Based Societies: Social and Cultural Interactions* (Idea Group Inc (IGI), 2009)
117. U. Visser et al., *RoboCup 2007: Robot Soccer World Cup XI*, vol. 5001 (Lecture Notes in Computer Science (Springer, Berlin, 2008)
118. I.A. Wagner, A.M. Bruckstein, Row straightening via local interactions, in *Technical report CIS-9406, Center for Intelligent Systems, Technion, Haifa* (1994)
119. I.A. Wagner, A.M. Bruckstein, Ants: agents, networks, trees and subgraphs. Future Gener. Comput. Syst. J. **16**(8), 915–926 (2000)
120. I.A. Wagner, A.M. Bruckstein, From ants to a(ge)nts: a special issue on ant–robotics. Ann. Math. Artif. Intell. Spec. Issue Ant Robot. **31**(1–4), 1–6 (2001)
121. I.A. Wagner, M. Lindenbaum, A.M. Bruckstein, Efficiently searching a graph by a smell-oriented vertex process. Ann. Math. Artif. Intell. **24**, 211–223 (1998)
122. I.A. Wagner, Y. Altshuler, V. Yanovski, A.M. Bruckstein, Cooperative cleaners: a study in ant robotics. Int. J. Robot. Res. (IJRR) **27**(1), 127–151 (2008)
123. P.K.C. Wang, Navigation strategies for multiple autonomous mobile robots, in *Proceedings of the IEEE/RSJ International Conference on Intelligent Robots and Systems (IROS)* (1989), pp. 486–493
124. A. Weitzenfeld, A prey catching and predator avoidance neural-schema architecture for single and multiple robots. J. Intell. Rob. Syst. **51**(2), 203–233 (2008)
125. M.P. Wellman, P.R. Wurman, Market-aware agents for a multiagent world. Robot. Auton. Syst. **24**, 115–125 (1998)
126. B.A. Whitmore, Evolution of unmanned aerial warfare: a historical look at remote airpower-a case study in innovation. Technical report, US Army Command and General Staff College Fort Leavenworth United States (2016)
127. H. Work, E. Chown, T. Hermans, J. Butterfield, Robust team-play in highly uncertain environments, in *AAMAS '08: Proceedings of the 7th international joint conference on Autonomous agents and multiagent systems* (International Foundation for Autonomous Agents and Multiagent Systems, Richland, SC, 2008), pp. 1199–1202
128. A. Yamashita, M. Fukuchi, J. Ota, T. Arai, H. Asama, Motion planning for cooperative transportation of a large object by multiple mobile robots in a 3d environment, in *In Proceedings of IEEE International Conference on Robotics and Automation* (2000), pp. 3144–3151
129. X.-S. Yang, Z. Cui, R. Xiao, A. Hossein Gandomi, M. Karamanoglu, *Swarm Intelligence and Bio-Inspired Computation: Theory and Applications* (Newnes, 2013)
130. V. Yanovski, I.A. Wagner, A.M. Bruckstein, Vertex-ants-walk: a robust method for efficient exploration of faulty graphs. Ann. Math. Artif. Intell. **31**(1–4), 99–112 (2001)
131. V. Yanovski, I.A. Wagner, A.M. Bruckstein, A distributed ant algorithm for efficiently patrolling a network. Algorithmica **37**, 165–186 (2003)
132. X. Zheng, S. Koenig, Reaction functions for task allocation to cooperative agents, in *Proceedings of the 7th International Joint Conference on Autonomous Agents and Multiagent Systems* (2008), pp. 559–566
133. R. Zlot, A. Stentz, M.B. Dias, S. Thayer, Multi-robot exploration controlled by a market economy, in *Proceedings of the IEEE International Conference on Robotics and Automation* (2002)

Cooperative *"Swarm Cleaning"* of Stationary Domains

1 Introduction

In this chapter we discuss the problem concerning a given (yet potentially unknown) region that needs to be *"cleaned"* by a swarm of autonomous robotic agents, namely – to have all of its 'tiles', or 'pixels' visited at least once by at least a single member of the swarm, and that this "cleaning" would be guaranteed to be completed within a finite (and as short as possible) time.

We assume that the swarm is collaborative, namely – that the various members of it are equipped with the same (and properly synchronized) software, and that each robot can acquire only the information which is available in its immediate vicinity, with the only way of inter-robot communication is by leaving traces on the common ground and sensing the traces left by other robots. Similar works can be found at [2, 21, 28, 29, 31].

In the spirit of [17], we consider simple robots with only a bounded amount of memory (i.e. *finite-state-machines*). We present a protocol for cleaning a dirty area that guarantees task completion (unless *all* robots die) and prove an upper bound on the time complexity of this protocol. Generalizing an idea from computer graphics (see [24]), we preserve the connectivity of the "dirty" region by allowing an agent to clean only so called *non-critical* points, points that do not disconnect the graph of dirty grid points. This ensures that the robots will stop only upon completing their mission.

An important advantage of this approach, in addition to the simplicity of the agents, is its fault-tolerance — even if almost all the agents cease to work before completion, the remaining ones will eventually complete the mission. We prove the correctness of the protocol as well as an upper bound on its running time, and show how our protocol can be extended for regions with obstacles. We also show simulation results of the protocol on several types of regions. These simulations indicate that

---

This chapter is based on work previously published in parts in [3, 9, 12].

© Springer International Publishing AG 2018
Y. Altshuler et al., *Swarms and Network Intelligence in Search*,
Studies in Computational Intelligence 729, DOI 10.1007/978-3-319-63604-7_2

the precise cleaning time depends on the number of robots, their initial locations, and the shape of the dirty region.

This chapter is organized as follows: in Sect. 2 the problem is formally defined, as well as the aims and assumptions involved. Related work is presented in Sect. 3. The cleaning protocol is presented in Sect. 4, while an analysis of its time-complexity appears in Sect. 5. Section 6 is devoted to the problem of exploring/cleaning regions with obstacles, and is followed by several experimental examples in Sect. 7. We conclude this chapter with a discussion and by pointing out several connections to related work, presented in Sect. 8.

2   The Cooperative Cleaners Problem

Throughout this work, we shall assume that the time is discrete.

Definition 1 Let G denote a two dimensional integer grid $\mathbf{Z}^2$, whose vertices have a binary property called *'dirtiness'*. Let $dirt_t(v)$ state the dirtiness state of the vertex v at time t, taking either the value *"on"* or *"off"*.

Definition 2 Let F_t be the dirty sub-graph of G at time t, i.e. $F_t = \{v \in G \mid dirt_t(v) = on\}$.

We assume that F_0 is a single simply-connected component (the actions of the agents will be so designed that this property will be preserved).

Let a group of k agents that can move on the grid G (moving from a tile to its neighbor in one time step) be placed at time t_0 on F_0, at point $p_0 \in F_t$.

Each agent is equipped with a sensor capable of telling the *dirtiness* status of all tiles in the digital sphere of diameter 7 (using the Manhattan distance function) which surrounds the agent. An agent is also aware of other agents which are located in these tiles, and all the agents agree on a common direction. Each tile may contain any number of agents simultaneously. This information will later be required by the agents' cleaning protocol. Each agent is equipped with a memory of size $O(1)$ bits w.r.t the size of the region.[1]

The agents are indistinguishable. Namely, they can be counted, but they do not contain any unique ID.

When an agent moves to a tile v, it has the possibility of cleaning this tile (i.e. causing $dirt_{(v)}$ to become *off*). The agents do not have any prior knowledge of the shape or size of the sub-graph F_0 except that it is a single and simply connected component.

The agents' goal is to clean G by eliminating the dirtiness entirely, meaning that the agents must ensure that:

$$\exists t_{success} \ s.t \ F_{t_{success}} = \emptyset$$

In addition, it is desired that time $t_{success}$ will be minimal.

---

[1] It should be stated that although the memory size of the agents is completely independent w.r.t to size of the problem, it is still $O(\log k)$ bits w.r.t to the number of agents (used for counting purposes).

In this work we impose the restriction of no central control and full 'de-centralization', i.e. all agents are identical and no explicit communication between the agents is allowed. An important advantage of this approach, in addition to the simplicity of the agents, is fault-tolerance — even if almost all the agents cease to work before completion, the remaining ones will eventually complete the mission, if possible.

3 Related Work

As mentioned in previous sections, the cooperative cleaners problem has significant similarity to other types of multi-agents problems, such as cooperative coverage, or cooperative de-mining problems. In recent years, a lot of effort went to designing systems and algorithms for handling such tasks. In this process, various models and assumptions concerning the agents and their capabilities were used.

In general, most of the techniques used for the task of a distributed coverage use some sort of cellular decomposition. For example, in [33] the area to be covered is divided between the agents based on their relative locations. In [18] a different decomposition method is being used, which is analytically shown to guarantee a complete coverage of the area. Another interesting work is presented in [1], discussing two methods for cooperative coverage (one probabilistic and the other based on an exact cellular decomposition). All of the works mentioned above, however, rely on the assumption that the cellular decomposition of the area is possible. This in turn, requires the use of memory resources, used for storing the dynamic map generated, the boundaries of the cells, etc. As the initial size and geometric features of the area are generally not assumed to be known in advance, agents equipped with merely a constant amount of memory will most likely not be able to use such algorithms. In this work, we are interested in a multi-agents system which could perform such a cooperative coverage with a use of minimal amount of memory, unrelated to the size and geometry of the covered (or cleaned) area (this requirement is presented in Sect. 2). Such a system is presented in the later part of this work, and its ability to guarantee a completion of the task is shown.

Surprisingly, while many existing works concerning distributed (and decentralized) coverage present analytic proofs for the completion of the task (e.g. [1, 16, 18]), unfortunately, most of them lack analytic bounds for the completion time (although in many cases an extensive amount of empirical results of this nature are made available by extensive simulations). Although a proof for the coverage completion is an essential key in the design of a multi-agents system, analytic indicators for its efficiency should also be sought for. In this work, such results, bounding the cleaning time of the agents, are presented in Sect. 5. An interesting work to mention in this scope is this of [26, 35], where a swarm of ant-like robots is used for repeatedly covering an unknown area, using a real time search method called *node counting*. By using this method, the robots are shown to be able to efficiently perform such a coverage mission, and analytic bounds for the coverage time are discussed.

A different approach to a similar problem is discussed in [14], where a swarm of collaborative drones uses a network-centric approach in order to maximize the interception probability of mobile targets, that are bound to move only on a pre-defined network of roads.

As most coverage problems focus on achieving a complete coverage (and sometimes — in the minimal time possible), it is worth mentioning in this scope the work of [19], in which an interesting additional constraint is added. As this work was designed for an autonomous painting system, the uniformity of the coverage was demanded (in order to maintain the same thickness of the paint layer). In addition, this work examined the problem of 3-D coverage, in contrary to most search/cover systems, usually discussing the 2-D version. However, the main focus of this work was the use of an efficient single agent for this mission, instead of a cooperative multi-agents one.

4 The Cleaning Protocol

In order to solve the *Cooperative Cleaners* problem we propose a cleaning protocol, called **SWEEP**. Generalizing an idea from computer graphics (presented in [24]), this protocol preserves the connectivity of the *dirty* region by preventing the agents from cleaning *critical points* — points which when cleaned disconnect the dirty region (see Sect. 4.1). This ensures that the agents stop only upon completing their mission.

At each time step, each agent cleans its current location (assuming it is not a critical point), and moves according to a local movement rule, creating the effect of a clockwise traversal along the boundary of the dirty region. As a result, the agents "peel" layers from the region, while preserving its connectivity, until the region is cleaned entirely. An illustration of two agents working according to the protocol can be seen in Fig. 1.

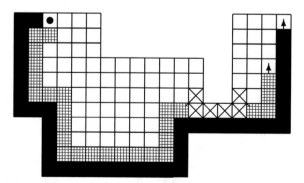

Fig. 1   An example of two agents using the **SWEEP** protocol, at time step 40. All the tiles presented were dirty at time 0. The *black dot* denotes the starting point of the agents. The X's mark the *critical points* which are not cleaned. The *black tiles* are the tiles cleaned by the first agent. The second layer of marked tiles represent the tiles cleaned by the second agent

To the basic description of the protocol given above, there are several exceptions. As the agents are equipped with very limited sensing and storage capabilities, the basic structure of the protocol must be enhanced with a set of local rules, designed for producing a pseudo-synchronization between the agents. Those rules generate *resting* and *waiting* commands for the agents, capable of delaying their actions, either within the time step (until certain agents complete their cleaning process for this time step), or causing them to pause for a single time step, resuming their operation at the next time step only. A detailed description concerning the need for these additions appears in Sects. 4.6 and 4.7. Note however that the rules in charge of the *resting* and *waiting* still obey the basic paradigm of this work, namely — they are local, use no prior knowledge of the agents, and can be implemented using extremely small amount of memory resources (namely —$O(\log k)$, k being the number of agents). A schematic flowchart of the protocol is presented in Fig. 3.

4.1 Cleaning Protocol — Definitions and Requirements

Definition 3 Let $\tau(t) = \big(\tau_1(t), \tau_2(t), \ldots, \tau_k(t)\big)$ denote the locations of the k agents at time t.

Definition 4 For a tile v, let $Neighborhood(v)$ denote the dirtiness states of v and its $8Neighbors$.

Let $\mathcal{M}_i$ denote some finite amount of memory contained in agent i, storing information needed for the protocol (e.g. the last moves of agent i). The requested cleaning protocol is therefore a rule f such that for every agent i:

$$f\big(\tau_i(t), Neighborhood(\tau_i(t)), \mathcal{M}_i(t)\big) \in \mathcal{D}$$

where $\mathcal{D} = \{`left`, `right`, `up`, `down`\}$.

Definition 5 Let ∂F denote the boundary of F. A tile is on the boundary if and only if at least one of its $8Neighbors$ is not in F, meaning:

$$\partial F = \{v \mid v \in F \ \wedge \ 8Neighbors(v) \cap (G \setminus F) \neq \emptyset\}$$

The requested rule f should meet the following goals:

- **Successful Termination**: $(\exists t_{success} \ s.t \ F_{t_{success}} = \emptyset)$.
- **Agreement on Completion**: within a finite time after completing the mission, all the agents must halt.
- **Efficiency**: the cleaning process should be efficient at time and in agents' memory resources.
- **Fault Tolerance**: if one or several stop working ("die") the rest of the agents will continue the cleaning process as efficiently as their number allows them.

4.2   The SWEEP Cleaning Protocol

The **SWEEP** protocol is implemented by each agent a_i, located at time t at $\tau_i(t) = (x, y)$. We define below several terms we use while discussing the **SWEEP** protocol. We stress the fact that this is indeed a *myopic* protocol, relying on neighborhood information, a *senile* protocol (the memory needed is constant + one counter whose size is upper bounded by $O(\log k)$, where k is the number of agents) and relies on *implicit local communication* only.

Definition 6   Let $\tilde{\tau}_i(t)$ denote the previous location of agent i with respect to $\tau_i(t)$, such that $\tilde{\tau}_i(t) \neq \tau_i(t)$, defined as:

$$\tilde{\tau}_i(t) \triangleq \tau_i(x) \; s.t. \; x = \max\{j \in \mathbb{N} \mid j < t \text{ and } \tau_i(j) \neq \tau_i(t)\}$$

Definition 7   The term *'rightmost'* means:

Starting from $\tilde{\tau}_i(t)$ (namely, the previous boundary tile that agent a_i had been in) scan the *four neighbors* of $\tau_i(t)$ in a clockwise order until a boundary tile (excluding $\tilde{\tau}_i(t)$) is found. Sometimes, $\tilde{\tau}_i(t)$ might not be a boundary tile (this may occasionally happen after a contamination spread, for dirty regions of specific geometric features. In those cases, an agent may find itself in an *internal point* — a dirty tile which is not located on the boundary. From this tile the agent must move to the adjacent boundary tile, and resume its traversal along the region's boundary). In this case, if $\tau_i(t)$ is a boundary point, then starting from $\tilde{\tau}_i(t)$ scan the *four neighbors* of $\tau_i(t)$ in a clockwise order until the second boundary tile is found. In case $t = 0$ select the tile as instructed in Fig. 2.

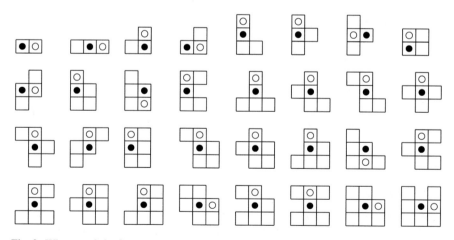

Fig. 2   When $t = 0$ the first movement of an agent located in (x, y) should be decided according to initial contamination status of the neighbors of (x, y), as appears in these charts — the agent's initial location is marked with a *filled circle* while the destination is marked with an *empty one*. All configurations which do not appear in these charts can be obtained by using rotations. This definition is needed in order to initialize the traversal behavior of the agents in the correct direction

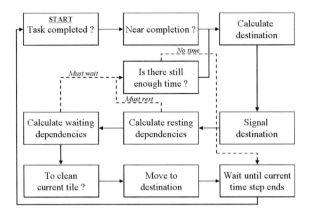

Fig. 3 A schematic flow chart of the **SWEEP** protocol. The *smooth lines* represent the basic flow of the protocol while the *dashed lines* represent cases in which the flow is interrupted. Such interruptions occur when an agent calculates that it must not move until other agents do so (either as a result of *waiting* or *resting* dependencies — see Lines 6 and 19 of **SWEEP** for more details)

The reference to *contamination spread* used in Definition 7 is given with consideration to the use of this term in the next chapters.

The additional information needed for the protocol and its sub-routines is contained in $\mathcal{M}_i$ and $Neighborhood(x, y)$.

A schematic flowchart of the protocol, describing its major components and procedures is presented in Fig. 3. The complete pseudo-code of the protocol and its sub-routines appears in Figs. 4 and 5. Upon initialization of the system, the *System Initialization* procedure is called (defined in Fig. 4). This procedure sets various initial values of the agents, and call the protocol's main procedure — *SWEEP* (defined in Fig. 5). This procedure in turn, uses various sub-routines and functions, all defined in Fig. 4.

4.3 Pivot Point

The initial location of the agents (the *pivot point*, denoted as p_0) is artificially set to be critical during the execution, hence it is also guaranteed to be the last point cleaned. Completion of the mission can therefore be concluded from the fact that all (working) agents are back at p_0 with no contaminated neighbors to go to, thereby reporting on completion of their individual missions. Note that this artificial preservation of the criticalness of the pivot point is not necessary for the algorithm to work. Rather, it just makes life easier for the user, as one can now know where to find the agents after the cleaning has been completed. If we do not start with all agents at the pivot and force p_0 to be critical, the location of the agents upon completion will generally not be known in advance.

```
1:  System Initialization
2:      Arbitrarily choose a pivot point p_0 in ∂F_0, and mark it as critical point
3:      Place all the agents on p_0
4:      For (i = 1; i ≤ k; i + +) do
5:          Call Agent Reset for agent i
6:          Call SWEEP for agent i
7:          Wait two time steps
8:      End for
9:  End procedure

10: Agent Reset
11:     resting ← false
12:     dest ← null /* destination */
13:     near completion ← false
14:     saturated perimeter ← false
15:     waiting ← ∅
16: End procedure

17: Priority
18:     /* Assuming the agent moved from (x_0, y_0) to (x_1, y_1) */
19:     priority ← 2(x_1 − x_0) + (y_1 − y_0)
20: End procedure

21: Check Completion of Mission
22:     If ((x, y) = p_0) and (x, y) has no dirty neighbors then
23:         If (x, y) is dirty then
24:             Clean (x, y)
25:         STOP
26: End procedure

27: Check "Near Completion" of Mission
28:     /* Cases where every tile in F_t contains at least a single agent */
29:     near completion ← false
30:     If each of the dirty neighbors of (x, y) contains at least one agent then
31:         near completion ← true
32:     If each of the dirty neighbors of (x, y) has near completion = true then
33:         Clean (x, y) and STOP
34:     /* Cases where every non-critical tile in ∂F_t contains at least 2 agents */
35:     saturated perimeter ← false
36:     If ((x, y) ∈ ∂F_t) and both (x, y) and all of its non-critical neighbors
        in ∂F_t contain at least two agents then
37:         saturated perimeter ← true
38:     If ((x, y) ∈ ∂F_t) and both (x, y) and all of its neighbors in ∂F_t has
        saturated perimeter = true then
39:         Ignore resting commands for this time step
40: End procedure
```

Fig. 4 The first part of the **SWEEP** cleaning protocol. The terms ∂F and $\tau_i = (x, y)$ are defined in Sects. 4.1 and 4.2

4.4 Signaling

Since by assumption the agent's sensors can detect the status of all tiles which are contained within a digital sphere of radius four placed around the current location of the agent, each agent can artificially calculate the desired destination of all the agents

1: **SWEEP Protocol** /* Controls agent i after **Agent Reset** */
2:      **Check Completion of Mission**
3:      **Check "Near Completion" of Mission**
4:      $dest \leftarrow$ *rightmost neighbor* of (x,y) /* Calculate destination */
5:      *destination signal bits* $\leftarrow dest$ /* Signaling the desired destination */
6:      /* Calculate resting dependencies (solves agents' clustering problem) */
7:      Let all agents in (x,y) except agent i be divided to the following groups :
8:          A_1 : Agents signaling towards any direction different than $dest$
9:          A_2 : Agents signaling towards $dest$ which entered (x,y) before agent i
10:         A_3 : Agents signaling towards $dest$ which entered (x,y) after agent i
11:         A_4 : Agents signaling towards $dest$ which entered (x,y) with agent i
12:     Let group A_4 be divided into the following two groups :
13:         A_{4a} : Agents with lower *priority* than this of agent i
14:         A_{4b} : Agents with higher *priority* than this of agent i
15:     $resting \leftarrow false$
16:     **If** $(A_2 \neq \emptyset)$ or $(A_{4b} \neq \emptyset)$ then
17:         $resting \leftarrow true$
18:         **If** (current time-step $\mathcal{T}$ did not end yet) then jump to 4 **Else** jump to 35
19:     $waiting \leftarrow \emptyset$ /* Waiting dependencies (agents synchronization) */
20:     Let *active agent* denote a *non-resting* agent which didn't move in $\mathcal{T}$ yet
21:     **If** (x-1,y) $\in F_t$ contains an active agent then $waiting \leftarrow waiting \cup \{left\}$
22:     **If** (x,y-1) $\in F_t$ contains an active agent then $waiting \leftarrow waiting \cup \{down\}$
23:     **If** (x-1,y-1) $\in F_t$ contains an active agent then $waiting \leftarrow waiting \cup \{l\text{-}d\}$
24:     **If** (x+1,y-1) $\in F_t$ contains an active agent then $waiting \leftarrow waiting \cup \{r\text{-}d\}$
25:     **If** $dest$ = right and (x+1,y) contains an active agent j, and $dest_j \neq left$, and
         there are no other agents delayed by agent i (i.e. (x-1,y) does not contain
         active agent l with $dest_l$ =right and no active agents in (x,y+1),(x+1,y+1),
         (x-1,y+1), and (x+1,y) does not contain active agent n with $dest_n = left$),
         then ($waiting \leftarrow waiting \cup \{right\}$) and $\left(waiting_j \leftarrow waiting_j \setminus \{left\} \right)$

26:     **If** $dest$ = up and (x,y+1) contains an active agent j, and $dest_j \neq down$, and
         there are no other agents delayed by agent i (i.e. (x,y-1) does not contain
         active agent l with $dest_l$ =up and no active agents in (x+1,y),(x+1,y+1),
         (x-1,y+1), and (x,y+1) does not contain active agent n with $dest_n = down$),
         then ($waiting \leftarrow waiting \cup \{up\}$) and $\left(waiting_j \leftarrow waiting_j \setminus \{down\} \right)$

27:     **If** $(waiting \neq \emptyset)$ then
28:         **If** ($\mathcal{T}$ has not ended yet) then jump to 4 **Else** jump to 35
29:     /* Decide whether or not (x,y) should be cleaned */
30:     **If** $\neg$ ((x,y) $\in \partial F_t$) or ((x,y) $\equiv p_0$) or (x,y) has 2 dirty tiles in its $4Neighbors$
         which are not connected via a path of dirty tiles from its $8Neighbors$ then
31:         (x,y) is an *internal point* or a *critical point* and should not be cleaned
32:     **Else**
33:         Clean (x,y) if and only if it does not still contain other agents
34:     Move to $dest$
35:     Wait until $\mathcal{T}$ ends.
36:     Return to 2

Fig. 5 The **SWEEP** cleaning protocol. The term *rightmost neighbor* is defined in Sect. 4.2. *l-d* and *r-d* are left-down and right-down, respectively

which are located in one of its $4Neighbors$ tiles (see Fig. 6). Thus, the *signaling* action of each agent can be simulated by the other agents near him, and hence an explicit signaling by the agents is not actually required. However, the signaling action is kept in the description and flowchart of the protocol (in procedure 5) for the sake of simplicity and understandability.

Fig. 6 Digital sphere of
diameter 7, placed around
the agent

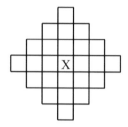

4.5 Connectivity Preservation

The connectivity of the region yet to be cleaned, F, is preserved by allowing the agents to clean only *non-critical* points. This guarantees both successful termination and agreement of completion (since having no dirty neighbors implies that $F = \emptyset$). Also, should several agents malfunction and halt, as long as there are still functioning agents, the mission will still be carried out, albeit slower.

Note that a tile which was originally clean, or which was cleaned by the agents, is guaranteed to remain clean. As a tile can be cleaned by the agents only when it is in ∂F, it is easy to see that the simple connectivity of F is maintained throughout the cleaning process (as the creation of "holes" — clean tiles which are surrounded by tiles of F — is impossible).

Note that when several agents are located in the same tile, only the last one who exits cleans it (see Line 33 in **SWEEP**), in order to prevent agents from being 'thrown' out of the dirty region (meaning, cleaning a tile in which another agent is located and thus preventing this agent from being able to continue the execution of the cleaning protocol). An alternative method for ensuring the agents are always capable of executing the cleaning protocol would have been the implementation of a "dirtiness seeking mechanism" (i.e. by applying methods such as suggested in [15]). Such a mechanism would have allowed an agent to clean its current location, even if other agents had also been present there. That solution, however, would have been far less elegant and would have added additional difficulties to the analysis process.

4.6 Agents Synchronization

Note that agents operating in the described environment must have some mean of synchronization, which is mandatory in order to prevent agents from operating in the same time — risking to cut the dirty region into several connected components, as shown in Fig. 7.

To ensure such scenarios will not occur, an order between the operating agents must be implemented. Note — throughout the next paragraphs agents which are signaling a *resting* status (see Sect. 4.7 for more details) are being disregarded while calculating the dependencies of the agents' movements. The creation of the following order is implemented in Procedure 19 of **SWEEP**:

Fig. 7 When the agents do not possess a synchronization mechanism, they may, among others, damage the region's connectivity. In this example, two agents clean their current locations, and move according to the **SWEEP** protocol. Since they are not synchronized, the tiles which they are located in are not treated as critical at the time of cleaning. However, the region's connectivity is not preserved. Should one of the agents had waited for its neighbor to complete executing the protocol's steps before resuming its actions, while deciding whether to clean its current location, it would have treated this tile as critical, and therefore avoid cleaning it. In this case, the connectivity of the region would have been maintained

Definition 8 For agent i, let $P_i \subseteq \{up, down, left, right, right\text{-}down, left\text{-}down\}$ be a set of directions of tiles, in which there are currently agents, which agent i is delayed by (meaning, agent i will not start moving until the agents in these tiles move). Unless stated otherwise, $P_i = \emptyset$.

For agent i which is located in tile (x, y), if $(x - 1, y)$ is a tile in F, which contains an agent, then $P_i \leftarrow P_i \cup \{left\}$. If $(x, y - 1)$ is a tile in F, which contains an agent, then $P_i \leftarrow P_i \cup \{down\}$, and similarly for $(x + 1, y - 1)$ and $right\text{-}down$ and for $(x - 1, y - 1)$ and $left\text{-}down$.

Definition 9 Let $dest_i \in \{up, down, left, right\}$ be the destination agent i might be interested in moving to, after leaving its current location.

Let each agent be equipped with a two bit flag (that may be implemented for example by two small light-bulbs). This flag states the desired destination of the agent. Alternatively, each tile can be treated as a physical tile, in which the agent can move. Thus, the agent can move towards the top side of the tile, which will be equivalent for using the flag in order to signal that it intends to move up.

Let each time step be divided into two phases. In phase 1, every agent "signals" the destination it intends to move towards, either by moving to the appropriate side of the tile, or by using the destination flag.

As we defined an artificial rule which states the superiority of $left$ and $down$ over $right$ and up (and internally, of $down$ over $left$), there are several specific scenarios in which this asymmetry should be reversed in order to ensure a proper operation of the agents. Following is such a "dependencies switching" rule: For agent i, which is located in (x, y), if $dest_i = right$ and tile $(x + 1, y)$ contains an agent j, and $dest_j \neq left$, and there are no other agents which are delayed by agent i (i.e. tile $(x - 1, y)$ does not contain an agent l where $dest_l = right$ and tiles $(x, y + 1)$, $(x + 1, y + 1)$, $(x - 1, y + 1)$ do not contain any agent and tile $(x + 1, y)$ does not contain an agent n where $dest_n = left$), then $P_i \leftarrow P_i \cup \{right\}$ and $P_j \leftarrow P_j \setminus \{left\}$. Also, if $dest_i = up$ and tile $(x, y + 1)$ contains an agent k, and $dest_k \neq down$, and there are no other agents which are delayed by agent i (i.e. tile $(x, y - 1)$ does not contain an agent m where $dest_m = up$ and tiles $(x + 1, y), (x + 1, y + 1), (x - 1, y + 1)$ do not contain any agent and tile $(x, y + 1)$ does not contain an agent q where $dest_q = down$), then $P_i \leftarrow P_i \cup \{up\}$ and $P_k \leftarrow P_k \setminus \{down\}$.

At phase 2 of each time step, the agents start to operate in turns, according to the order implied by P_i. This guarantees that the connectivity of the region is kept, since the simultaneous movement of two neighboring agents is prevented.

Notice that deadlocks are impossible — since the basic rule is that every agent is delayed by the agents in its *left* and *down* neighbor tiles. Therefore, at any given time, and for every possible group of agents, there exist an agent with the minimal x and y coordinates (which by definition, is not delayed by any other agent of this group). After this agent moves, all the agents which are delayed by it, can now move, and so on. As to the "dependencies switching rule" — let agent i located in tile (x, y) have the minimal x and y values among the agents who had not moved yet, and let $dest_i = up$, and let tile $(x, y + 1)$ contain an agent j such that $dest_j \neq down$. Then although agent i is located below agent j, it will be delayed by it (i.e. $(up \in P_i)$ and $\neq (down \in P_j)$) as long as agent i is not delaying any other agent (as this is the requirement of the "dependencies switching rule"). In this case, we should show that there can not be a cycle of dependencies, which starts at agent j, ends at agent i, and is closed by the dependency of agent i on agent j. Such a circle can not exist since for it to end in agent i, it means that agent i is delaying another agent k. However, this is impossible since agent i is known not to delay any agent (specifically agent k). Hence, circular dependencies are prevented and no deadlock are possible.

Note that while phase 2 is in process, F_t may change due to cleaning by the agents. As a result, the desired destinations of the agents as well as their dependencies forest, must also be dynamic. This is achieved through the repeated recalculation of these values, for every *waiting* agent. For example, assume agent i to be located in (x, y), and $dest_i = down$ without loss of generality, and let agent j located in tile $(x, y - 1)$ moves out of this tile and cleans it. Then, $dest_i$ naturally change (as the tile (x, y) does no longer belong to F_t, and thus is not a legitimate destination for agent i), and thus, P_i may also change. In this case, agent i should change its "destination signal" and act according to its new $dest_i$ and P_i. This is implemented in **SWEEP** by calculating the waiting agents' destinations and dependencies lists repeatedly, until either all agents have moved or until the time step has ended (meaning that some agents had to change their status to *resting*, and pause until the next time step — see Sect. 4.7 for more details). Note that every *waiting* agent is guaranteed either to complete its movement in the current time step, or to be forced to wait for the next time step, by switching its status to *resting*.

Notice that the "dependencies switching rule" is not required in order to ensure a proper completion of the mission, but rather to improve the agents' performance, by preventing a bug demonstrated in Fig. 8.

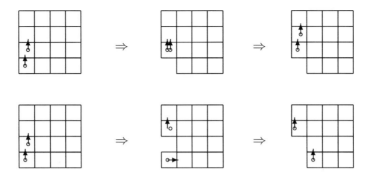

Fig. 8 The *upper charts* demonstrate a performance bug which may be caused due to local dependencies. The agents are advancing according to the **SWEEP** protocol, but their cleaning performance is decreased. The *lower charts* demonstrates the cleaning operation after adding the dependencies switching mechanism

4.7 Clustering Problem

Since we are interested in preventing the agents from moving together and functioning as a single agent (due to a resulting decrease in the system's performance), we would like to impose a "resting policy" which will ensure that the agents do not group into clusters which move together. This is done by artificially delaying agents in their current locations, should several agents placed in the same tile wish to move to the same destination. However, we would like the resting time of the agents to be minimal, for performance and analysis reasons.

The following resting policy is implemented by Line 6 of **SWEEP**. Using its sensors, an agent intending to move into tile v is aware of other agents which intend to move into v as well. Thus, when an agent enters a tile which already contains other agents, it knows whether these agents entered this tile at the same time step it did, or whether they had occupied this tile before the current time step had started.

Note that in phase 1 of each time step all the agents are signaling their desired destinations. Thus, an agent can identify which ones of the agents which are located in its current tile intend to move to the same destination as its. Only those agents may cause this agent to delay its actions and rest.

From the agents which intend to move to the same direction as agent i, the agent can distinguish between three groups of agents — the agents which entered this tile before the time step agent i did (group A_2), those which entered the tile in the same time step as i (group A_4), and those who entered it after agent i did (group A_3). If $A_2 \neq \emptyset$, agent i waits, change its status to *resting agent*, signals this status to the other agents in its vicinity, and does not move. As this rule is kept by every other agent, agent i is guaranteed that the agents of group A_3 in their turn will wait for agent i moves before being free to do so as well.

As to the agents of A_4, the problem is solved by induction over the time steps, namely — that at any given time, two agents can not leave the same tile towards the same direction at the same time step. The base of this induction holds for small values of t as the agents are periodically released by the initialization procedure of **SWEEP**, while no two agents are released at the same time step. Later, as we are assured that all the agents of group A_4 arrived to v from different tiles (which are also different than the tile agent i had entered v from), then since all the agents in group A_4 know the previous locations of each other, a consensus over a local ordering of A_4 is established (note that no explicit communication is needed to form this ordering). An example for such an order is the **Priority** function of the **SWEEP** protocol. As a result, the agents are able to exit the tile they are currently located in, in an orderly fashion, according to a well defined order. Thus, the following invariant holds: *"at any given time t, for any two tiles v, u, there can only be a single agent which moves from v to u at time step t"*. Thus, the clustering problem is solved.

Notice that there is a single exception to this mechanism, in which the resting commands are overruled. This happens when all non-critical perimeter tiles contain at least two agents. In this scenario, following the resting commands would have created a situation in which no tile can be cleaned, as for every agent which leaves a certain tile, there are still "resting" agents located in the same tile. Therefore, once this scenario is detected by the *check near completion of mission* procedure of **SWEEP**, the agents ignore their resting commands momentarily, in order to be able to clean the non-critical perimeter tiles. The order and internal prioritization of the agents are maintained, for calculating the new *resting* commands in the following time step.

4.8 Mission Termination

The termination of the protocol is done in one of two cases — either an agent finds itself in the pivot point, while all of its neighbors are clean (which means that this is the last dirty tile and therefore should be cleaned), or when each dirty tile contains at least one agent (which is a generalization of the previous scenario). The second case is implemented by allowing the agents to signal to their four neighbors whether all of their dirty neighbors contain at least a single agent.

5   Analysis

The introduction of the notion of critical points makes time-complexity analysis of the **SWEEP** protocol significantly harder since a critical point may be visited several times before it is cleaned. We conjecture that the protocol proposed is efficient, and additional agents will speed up the cleaning process only up to a limit. In order to give this argument a rigorous form, some definitions are required.

5.1 Definitions

Definition 10 Let S_t denote the size of the dirty region F at time t, namely the number of grid points (or tiles) in F_t.

Actually, F defines a dichotomy of $\mathbf{Z}^2$ into F and $\overline{F} = \mathbf{Z}^2 \setminus F$.

Recalling Definition 5, the boundary of the dirty region F is denoted as ∂F and defined as:

$$\partial F = \{(x, y) \mid (x, y) \in F \wedge (x, y) \text{ has an } 8Neighbor \text{ in } (G \setminus F)\}$$

Definition 11 A *path* in F is defined to be a sequence $(v_0, v_1, \ldots, v_n)$ of tiles in F such that every two consecutive tiles are 4 *connected* (the *Manhattan* distance between them is 1). The *length* of a *path* is defined to be the number of tiles in it.

Definition 12 Let tile v be called a *critical point* if there exist $v_1, v_2 \in 4$ $Neighbors(v)$ for which all paths connecting v_1 and v_2, included in $8Neighbors(v)$, necessarily pass through v (where v, v_1, v_2 and all said paths are in F).

Definition 13 We shall denote by c_t the circumference of F at time t, defined as follows: let v_0 and v_n be two 4 *connected* tiles in ∂F_t, and let $C_t = (v_0, v_1, \ldots, v_n)$ be a shortest *path* connecting v_0 and v_n, which contains all the tiles of ∂F_t and only such tiles. Notice that C_t may contain several instances of the same tile, if this tile is a *critical point* (meaning that C_t is an ordering of the tiles of ∂F_t, in which multiple instances of tiles that are *critical points* are allowed). c_t will be defined as the length of C_t. An example appears in Fig. 9.

Fig. 9 The *line* in the *left upper chart* goes through the tiles of C_t where the 3 *arrows* denote tiles that are included twice. The *circles* in the *right upper chart* denote the tiles of ∂F_t. Note that while ∂F_t contains 11 tiles, C_t contains 14. In the *lower chart*, there are no *critical points* in ∂F_t and therefore $c_t = |\partial F_t| = |C_t| = 36$

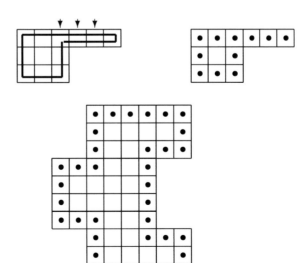

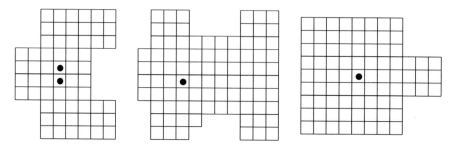

Fig. 10 The chart contains three regions. The *black circles* denote the deepest tiles within the regions. Notice that while in the left region there are two tiles whose depth is the maximal, in the other regions there is only one tile of maximal depth. The widths of the regions equal the depths of the deepest tiles, and are (from *left* to *right*) 2, 3, and 4

Definition 14 For some $v \in F_t$ let $Strings_t(v)$ denote the set of all *paths* in F_t that begin in v and end at any *non-critical point* in ∂F_t, and let $w(F_t, v)$ denote the *depth* of v — the length of the shortest *path* in $Strings_t(v)$ (unless v is a *critical point* in which case its *depth* is defined to be zero).

Definition 15 Let $W(F_t)$ denote the *width* of F_t, defined as the maximal depth of all the tiles in F_t, i.e.:

$$W(F_t) = \max\{w(F_t, v) \mid v \in F_t\}$$

An example appears in Fig. 10.

Definition 16 Let F_t' denote the region one could get, having one agent traverse F_t once, using the **SWEEP** protocol (i.e. when $k = 1$, $F_t' = F_{t+c_t}$). For the sake of simplicity, F_t'' will be used instead of $(F_t')'$, and so on.

Definition 17 The longest in-region distance between p_0 and any other point in F_t will be referred to as $l(F_t)$, the *length* of F_t:

$$l(F_t) = \max_{v \in F_t} \left\{ d_{F_t}(p_0, v) \right\}$$

where $d_{F_t}(x, y)$ is the distance (shortest path within F_t) between x and y.

5.2   Correctness

Lemma 1 *If $S_t > 0$ then $W(F_t) \geq 1$, that is: any finite simple region has at least two non-critical points on its boundary.*

Proof The boundary ∂F_t is a connected graph, hence it has a spanning tree. In such a tree there are at least two vertices with degree 1. These vertices are necessarily not critical in F_t, since any boundary point which is critical in F_t is also critical in ∂F_t.

Lemma 2 *During the course of the cleaning process, if all the agents are using the* **SWEEP** *protocol the agents are never going outside the dirty region. Namely, the only time an agent is located on a clean tile is when* $F = \emptyset$.

Proof The only movements allowed by the **SWEEP** protocol are towards a dirty tile. The only times an agent is allowed to clean a tile are when it is the last agent leaving this tile, or when cleaning this tile means that the mission is completed (for example, when this is the only dirty tile left). In addition, the various preemption which are in charge of agents synchronization prevent an agent to move to a dirty tile, which is getting cleaned during its movement. As a result, an agent is guaranteed to be located in dirty tiles throughout the entire cleaning process.

Theorem 1 *A group of agents executing the* **SWEEP** *protocol will eventually clean a simply connected region and stop together at the pivot point* p_0.

Proof While F_t has not yet been cleaned, $S_t > 0$ and hence, by Lemma 1 there is a non-critical point on ∂F_t. Since the agents obey the **SWEEP** protocol, while $F_t \neq \emptyset$, there is at least a single agent still traversing F_t. As this agent does not change the direction of its traversal, it is guaranteed to arrive to (and clean) a non-critical point at least once during the course of its traversal (thus decreasing S_t by at least 1). As $c_t \leq 4S_t$ (since during a traversal around F_t any tile can be entered to from each one of its 4 neighbors at most once), the maximal length of this traversal is $4S_t$. As agents may indeed delay one another, it is shown in Lemma 5 that the actual time required for an agent to complete its traversal around F_t is at most four times the length of this traversal, namely — $16S_t$. As $\forall t$, $S_t \leq S_0$, it can be seen that after no more than $16s_0^2$ time steps we shall have $S_t = 0$. The fact that all agents will meet at the same point is implied by the following two rules that are implemented in the **SWEEP** protocol:

- **Rule 1**: F_t is always being kept connected (we never clean a tile that has no clean neighbor and never clean a critical tile either).
- **Rule 2**: The pivot p_0 is cleaned only at the last time step of the cleaning process.

5.3 Detailed Analysis

In this section, we shall discuss an upper bound for the cleaning time of k agents employing the **SWEEP** cleaning protocol, for some dirty region F_0. From an upper bound for the cleaning time of a swarm comprising k agents, an upper bound for the number of agents needed to ensure a successful cleaning of a given region in a given number of time-steps can be derived.

The following Lemma states the change in the cardinality of a region's circumference after being traversed by an agent using the **SWEEP** protocol.

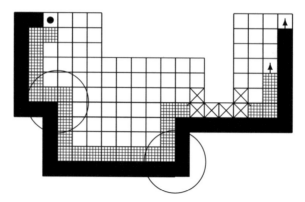

Fig. 11 An illustration of Lemma 3 and of the **SWEEP** protocol. The *black dot* denotes the starting point of the agents cleaning F (which is the entire region presented). The X's mark the *critical points* which are not cleaned. The *black tiles* are some of the tiles of ∂F, cleaned by the first agent. The second layer of marked tiles represent some of the tiles of $\partial F'$, cleaned by the second agent. The effect of corners on the traversal cardinality is either an increase (marked by the *left circle*) or a decrease (marked by the *right circle*)

Lemma 3 *The cardinality of the region's circumference always decreases after being traversed by an agent applying the* **SWEEP** *protocol (assuming no spread occurs), namely:*

$$|\partial F'| \leq \max\{1, |\partial F| - 8\}$$

Proof For any region F such that the number of tiles in F' is at least 2, note that a traversal around F is a closed, simple, rectilinear polygon. $\partial F'$ was obtained after deleting all the *non-critical* points of ∂F. Traversing such a polygon, an agent either goes straight, makes an internal turn ("left" turn, if we assume a clockwise movement), or makes an external turn ("right" turn). Suppose without loss of generality that an agent is moving up and making an internal turn left. Assume that there are no critical points. The path around this turn, along the newly created boundary tiles is now longer by two, as it contains an additional movement up, and another one towards the left. Similarly, an external turn, creates a new path which is shorter by two tile (as a single horizontal movement and a single vertical movement are no longer necessary). Since ∂F is a simple rectilinear polygon, it always has four "right" turns more than "left" ones.[2] When *critical* points are met, they are not being cleaned, which means that they exist in both ∂F and $\partial F'$. Re-visiting these points, however, does not change the overall size of the set of tiles (see an example in Fig. 11).

This proof, though, does not always hold for regions that, after being traversed once, produce either $\emptyset$ or a single tile. The reason is that the proof of the Lemma that appears in [36] relies on the fact that for each "turn" in F, there is a corresponding

---

[2]This is a simple consequence of the "rotation index" Theorem (see e.g. [20] p. 396): If $\alpha : [0, 1] \to R^2$ is a plane, regular, simple, closed curve, then $\int_0^1 k(s)ds = 2\pi$, where $k(s)$ is the curvature of $\alpha(s)$ and the curve is traversed in the positive direction.

"turn" in F'. However, the concept of "turn" requires the existence of a possible movement of an agent from one tile to another time. This obviously cannot be done for $F' = \emptyset$ or for $F' = \{v\}$. For such regions it can easily be seen the Lemma holds, as $|\partial\emptyset| = 0$ and as $|\partial\{v\}| = 1$.

While producing a bound for the cleaning time of the agents we shall demonstrate that while being traversed by the agents using the **SWEEP** protocol, the width of the initial region F_0 decreases monotonically. The main component of this proof is presented in the following Lemma.

Lemma 4 *Every time a region is traversed by an agent using the **SWEEP** protocol, its width is decreased by at least one, namely:*

$$W(F'_t) \leq W(F_t) - 1$$

A corollary is that the number of tours around a region F that k agents must accomplish before $F = \emptyset$ is at most $(\frac{W(F)}{k} + 1)$.

Proof By the definition of the depth of a tile, it is the shortest path to a *non-critical* tile in ∂F. According to the **SWEEP** protocol, after an agent had traversed F, all of the *non-critical points* in ∂F were cleaned. This is true as the agent is instructed to clean all the *non-critical points* it is going through. The only exception are tiles which an agent does not clean upon exit, as there are other agents currently located in it. However, in this case, this tile will be cleaned by the last agent leaving it, at the following time steps. Therefore, after an agent traverses the region, the depth of every internal tile was decreased by (at least) one, meaning that the total width of the region was decreased by (at least) one. As a result, as all agents are identical, when k agents complete a traversal of F, the width of F is decreased by at least k.

When examining the cleaning time of the agents, one must remember that merely calculating the length of the paths the agents must move along is not enough. Since in various scenarios the **SWEEP** protocol may direct an agent to stop and wait (for example, when several agents are located in the same tile), a way to bound the actual time it takes an agent to move along a specific path must be found. The following Lemma establishes the relation between the length of a path and the maximal time it takes an agent using the **SWEEP** protocol to move along it.

Lemma 5 *The time it takes an agent which uses the **SWEEP** protocol to move along a path of length c_t (including delays caused by other agents located in the same tiles) is at most $4 \cdot c_t$.*

Proof According to the problem's specifications, each agent moves along the dirty region F at a pace of one tile per time step. The only exception may occur when several agents are delayed after entering the same tile.

According to the **SWEEP** protocol, although several agents can enter the same tile at the same time step, agents located in the same tile can leave it at the same time step only if they do so towards different directions. As a result, there arises the question whether more and more agents can enter a certain tile without leaving it, causing a decrease in the swarm's performance.

Let v be an empty tile. Let us assume without loss of generality that at some time step t, 4 agents enter v (this is of course the maximal number of agents which can enter v at the same time step, due to the invariant stating that no two agents can exit u to v at the same time step, for u, some neighbor of v, and since v has at most 4 neighbors). Thus, in time step $t - 1$ v had exactly 4 neighbors in ∂F. Let us assume that in time step $t + 1$ v still has 4 neighbors in ∂F. Thus, all the 4 agents which entered v intends to move to 4 different directions (since according to the **SWEEP** protocol, each of them has different *rightmost* neighbor). Thus, none of them will be delayed. Alternatively, let us assume that at time step $t + 1$ v has less than 4 neighbors in ∂F. Let α denote the number of neighbors of v in ∂F. Then, since the 4 agents which entered v in time t did so from different directions, according to the **SWEEP** protocol, α of them will be able to leave v towards the α neighbors of v, and $(4 - \alpha)$ will stay in v. However, in this time step, only α new agents can enter v, since v has only α neighbors in ∂F. Thus, in time step $t + 2$ there are at most 4 agents in v. This is true of course for each time steps $T > t + 2$. Since the number of agents in each tile is at most 4, and since each tile with agents in it has at least one neighbor of ∂F (since the agents had to enter it from a neighbor in ∂F), then there are at least $\frac{k}{4}$ agents which are able to move. Thus, even if the agents collide with each other continuously, the time it takes k agents to traverse a region of circumference c_t is at most $4 \cdot k \cdot c_t$ (and the average traversal time of an agent, amortized for the entire group of agents, is at most 4 times its minimal traversal time).

Notice that this still holds for the rare occasions in which the *check near completion of mission* procedure detects that every non-critical perimeter tile of F_t contains at least two agents. In his case, the *resting* commands are indeed overruled, allowing temporarily more than 4 agents to be located into the same tile, however — as the internal prioritization of the agents is maintained, this is corrected at the following time steps, as the "redundant" agents will wait (sometimes for more than 4 time steps). Notice though that this "long resting" was already compensated by the "untimely movement" those agents had performed while entering this tile.

It should be stated that in reality, the agents' traversal time is only slightly more than c_t, although an analytic proof is yet to be found.

When an agent using the **SWEEP** protocol is traversing F_t it does so by moving along a path which contains all the tiles of ∂F_t. Since by doing that, the agent may pass through the same tile more than once, the length of this path may be larger than the number of tiles in ∂F_t. A bound for the ratio between the two is described in the following Lemma.

Lemma 6 c_t, *the length of the circumference of F_t never exceeds twice its cardinality, namely:*

$$c_t \leq 2 \cdot |\partial F_t| - 2$$

Fig. 12 The recursive spanning tree construction algorithm. The term *rightmost* has the same meaning as in the **SWEEP** protocol. By checking L we can find out whether continuing to an **OLD** neighbor is a "clean return" (e.g. (A,B,C,B)) or whether it completes a circle (e.g. (A,B,C,A))

```
 1: Add (x, y) to L and mark it as OLD
 2: Let v denote the rightmost neighbor of (x, y)

 3: If v ≡ p_s then
 4:     Delete all the marked tiles from L
 5:     STOP
 6: End if

 7: If (v is marked as OLD) and (no circle was formed) then
 8:     /* The traversal repeats this tile as well */
 9:     Call SPAN-DFS for v
10:     Upon return — STOP
11: End if

12: If (v is marked as OLD) and (a circle was formed) then
13:     /* Continue, and clean the redundant tiles */
14:     Add to L the sequence of tiles from L, starting with (x, y)
        and backwards to v
15:     Mark these tiles (excluding (x, y)) to be deleted
16:     Call SPAN-DFS for v
17:     Upon return — STOP
18: End if

19: /* The rightmost neighbor was not OLD */
20: Call SPAN-DFS for v
21: Upon return — STOP
```

Proof ∂F_t is a connected graph and thus it has spanning trees. An algorithm for constructing a spanning tree for ∂F_t and a *Depth First Search* (DFS) scan of it was constructed, such that the path generated by using the DFS algorithm contains a traversal around F_t (meaning that the DFS scan generates a path which after having several of its tiles removed, equals a path which traverses F_t).

This recursive spanning tree construction algorithm appears in Fig. 12. The algorithm receives p_s, a *non-critical* tile as its starting point and an empty list of tiles, L.[3] The algorithm constructs a spanning tree of ∂F_t as well as a DFS scan for this tree, by adding tiles to the list, while optionally marking some of them as tiles which should later be deleted. After the deletion of these tiles the remaining tiles form a path which traverses F_t. An example of the above appears in Fig. 13.

A DFS scan of a tree can go through each edge at most twice. A tree of $|V|$ vertices contains exactly $(|V| - 1)$ edges. Thus, a DFS scan of a spanning tree of ∂F_t contains at most $2 \cdot (|\partial F_t| - 1)$ transitions of edges. Since there exists such a spanning tree that contains a 'tour' around F_t then: $c_t \leq 2 \cdot |\partial F_t| - 2$.

Definition 18 Let $W_{REMOVED}(t)$ denote the decrease in F's width due to the agents cleaning activity until time t.

---

[3]The existence of a *non-critical* point is guaranteed since ∂F_t is a connected graph and thus has a spanning tree, in which at least two tiles has a degree of 1, which makes them *non-critical* tiles.

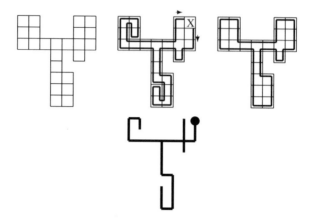

Fig. 13 An example of the spanning tree protocol. The *left chart* represents the dirty region F_t. The *middle chart* shows the DFS scan of ∂F_t according to the **SPAN-DFS** protocol, when the big X marks p_s, the *non-critical* starting point. The *right chart* shows the traversal path which is contained in this DFS scan. The drawing on the *bottom* shows the spanning tree that is created by the protocol and is searched by the DFS algorithm

Lemma 7

$$\text{If } \left\lfloor \sum_{i=1}^{t} \frac{k}{4 \cdot c_i} \right\rfloor \geq W(F_0) \text{ then } W_{REMOVED}(t) \geq W(F_0)$$

Proof $W_{REMOVED}(t)$ can be defined as the number of completed traversals around the dirty region, multiplied by the number of the agents k. Note that at time step t, each agent completes $\frac{1}{c_t}$ of the circumference. Since we know the value of c_t for every t, and since we know that the time it takes an agent to move along a path is at most 4 times the length of this path (see Lemma 5), then the decrease in the original width of the region caused by the cleaning process of the agents is bounded as described.

Note that as the expression above does not define between a multitude of agents working for a short while, and a single agent which is working for a long period of time, it may not hold for smaller values of t (in which the accumulated partial peelings is smaller than $W(F_0)$). Therefore, while merely writing $W_{REMOVED}(t) \geq \left\lfloor \sum_{i=1}^{t} \frac{k}{4 \cdot c_i} \right\rfloor$ might not always hold, the expression as appears above does.

A bound over the cleaning time of a given region F_0 can then be produced as follows:

Theorem 2 *Assume that k agents start cleaning a simple connected region F_0 at some boundary point p_0 and work according to the **SWEEP** protocol, and denote by $t_{success}(k)$ the time needed for this group to clean F_0. Then it holds that:*

$$\max \left\{ 2k, \left\lceil \frac{S_0}{k} \right\rceil, 2l(F_0) \right\} \leq t_{success}(k) \leq \frac{8(|\partial F_0| - 1) \cdot (W(F_0) + k)}{k} + 2k$$

Proof The lower bound is quite obvious — the left term $2k$ is the time needed to release the k agents from the pivot point, $\frac{S_0}{k}$ is a lower bound on the time necessary to cover the region if the agents were optimally located at the beginning. The right term, $2l(F_0)$, comes from the observation that at least one agent should visit (and return from) any point of F_0, including the one farthest from p_0, to which the distance is $l(F_0)$.

Considering the upper bound, in order for F_0 to be cleaned, there should exist a time $t_{success}$ in which the width of the region will be 0. Note that once the dirty region was reduced to a 'skeleton' comprises only of tiles of depth zero, an additional traversal must be done in order to clean it entirely.

By combining Lemmas 3 and 6 we know that for $c_j(i)$ and $\partial F_j(i)$ be the length and cardinality respectively of the i-th traversal of agent j:

$$c_j(i) \leq 2(|\partial F_j(i)| - 1) \leq 2(|\partial F_0| - 1)$$

Using Lemma 7 we see that:

$$W_{REMOVED}(t) \geq \left\lfloor \sum_{i=1}^{t} \frac{k}{4 \cdot c_i} \right\rfloor \geq \left\lfloor \frac{k \cdot t}{8(|\partial F_0| - 1)} \right\rfloor \tag{1}$$

Thus, we are interested in:

$$\left\lfloor \frac{k \cdot t_{success}(k)}{8(|\partial F_0| - 1)} \right\rfloor \geq W(F_0) + k \tag{2}$$

adding the "$+k$" for the additional traversal needed for cleaning the dirty 'skeleton'.

Since releasing the agents requires an additional $2k$ time steps, the final bound for this case is:

$$t_{success}(k) \geq \frac{8(|\partial F_0| - 1) \cdot (W(F_0) + k)}{k} + 2k$$

Using the above Theorem we can bound the *efficiency ratio*, defined as $\frac{t_{success}(k)}{S_0}$ which expresses the benefit of using k agents for a cleaning mission:

Corollary 1

$$\max \left\{ \frac{2k}{S_0}, \frac{1}{k}, \frac{2l(F_0)}{S_0} \right\} \leq \frac{t_{success}(k)}{S_0} \leq \frac{8\,(|\partial F_0| - 1) \cdot (W(F_0) + k)}{k \cdot S_0} + \frac{2k}{S_0} \tag{3}$$

An interesting result of Corollary 1 is that when $W(F_0) \ll k \ll S_0$, i.e. the number of agents is large relative to the width but small compared to the area, then the efficiency ratio is bounded below by twice the ratio of the length and area, and bounded above by $8\frac{|\partial F_0|}{S_0}$. Note here the similarity to the ratio $\frac{c_0}{\sqrt{S_0}}$, known as the *shape-factor*.

Another conclusion is that when we scale up the region by a factor of n, the area increases as n^2 but the width, length and circumference all increase as n so we get the following:

Corollary 2

$$as \; n \to \infty, \qquad L_\infty \leq \frac{t_{success}(k)}{S} \leq U_\infty$$

where

$$L_\infty = \frac{1}{k}, \qquad U_\infty = 8\frac{\partial F \cdot W}{k \cdot S} = 8\frac{|\partial F_0| \cdot W(F_0)}{k \cdot S_0}$$

and $S = S_0 n^2$, $\partial F = |\partial F_0|n$, $W = W(F_0)n$ are the scaled area, circumference cardinality and width, respectively.

5.4 Robustness

The issue of an algorithm's robustness and fault-tolerance is a major aspect of robotics systems, as the environments in which the agents operate may often include unpredicted changed from the original specification of the system, such as various kinds of noises, robots' malfunctions, etc.

While analyzing the performance of the cleaning protocol, the robotic agents were assumed to work perfectly. In real robotic system, however, this is not always the case. In simulation, however, it was quite easy to examine what happens when one or more agents disappear. The disappearance of an agent which dies is necessary in our framework in order for the system to work properly.

As discussed in Sect. 4.1, one of the requirements of the **SWEEP** protocol is a complete fault-tolerance to malfunctions in the agents, meaning — even if one or several agents stop working ("die" and disappear) the rest of the agents will continue the cleaning process as efficiently as their number allows them. The basic idea enabling this robustness is the complete independence between the agents, due to the full decentralized nature of the system (as the agents do not share the cleaning mission between them, nor do they rely on one another in the performance of their cleaning process).

A different aspect of the algorithm's robustness is the performance of the agents in noisy environments. In general, random noise can appear and influence various aspects of the system — in sensing whether a tile should be cleaned, in sensing the presence of other agents in an agent's vicinity, in the agent's movement, etc. As in almost any system, it is easy to see that severe noise may significantly affect the algorithm's efficiency, and may even prevent the agents from completion the mission altogether. For example, difficulties in sensing whether a tile should be cleaned or not, may cause an agent to clean a critical point, thus separating the region into several connected components. In addition, noises in the agents' movement system may cause the agents to detach from the contaminated region, thus delaying, or even preventing the completion of the cleaning process. Whereas a situation in which the noise level is kept at a reasonable level (i.e. correct identification of clean and contaminated tiles, and movement noises), it is the authors' opinion that the completion of the algorithm may still be achieved, although the cleaning will increase. When the agent has difficulties in the sensors in charge of detecting other agents in its vicinity, and interpreting their signaling (see relevant sections for details about signaling and synchronization) several problematic scenarios (which these signaling mechanisms were designed to avoid), may occur. However, simulations show that in such cases often only the cleaning time is affected, while the completion of the cleaning mission is preserved. Nevertheless, as this issue was not fully investigated yet, a perfect neighbors detection is still one of the algorithm's requirements.

One example for the above described problem is presented Fig. 7, in which a failure in the sensing mechanism may damage the simple-connectivity of the contaminated region. Note that in such scenarios, there is least a single agent in each of the new contaminated components. Therefore, each component is an instance of the cooperative cleaners problem, for which the upper bound for the cleaning time may be applied.

Another example is demonstrated in Fig. 8, where a bug in the *waiting* mechanism (see Sect. 4.6 for more details) may prevent some of the agents from cleaning contaminated tiles. The same holds for possible bugs in the *resting* mechanism (see Sect. 4.7 for more details) which may cause several agents to move together, acting as a single cleaning agent.

It is also interesting to examine the scenario in which all the robots are picked up, and re-placed on some arbitrary positions onboard the dirty region. Thanks to the definition of the SWEEP protocol, the robots will navigate to the boundary of the region, and resume their cleaning. Naturally, the overall cleaning time might be affected, however — the completion of the cleaning is still guaranteed.

5.5 Comparison to Existing Work

Looking at the existing work that has been done with respect to multi robotic systems designed for a cooperative cleaning/coverage/etc., several analytic results concerning the completion time of the mission can be found.

An interesting result is presented in [35]. In this work, a group of simple robots is used for periodically covering an unknown area, using the nodes counting method. Based on a more general result for undirected domains shown in [27], the following bound is given:

The cover time of teams of ant robots (of a given size) that use node counting on strongly connected undirected graphs can be exponential in the square root of the number of vertices.

Namely:

$$f(k) = O(2^{\sqrt{S_0}}) \tag{4}$$

denoting the covering time of k robots by $f(k)$.

Recalling Theorem 2 in which:

$$t_{success}(k) \leq \frac{8(|\partial F_0| - 1) \cdot (W(F_0) + k)}{k} + 2k$$

as $|\partial F_0| = O(S_0)$ and as $W(F_0) = O(\sqrt{S_0})$ we see that:

$$t_{success}(k) = O\left(\frac{1}{k} S_0^{1.5} + S_0 + k\right)$$

As for practical reasons we assume that $k < S_0$ we can see that:

$$t_{success}(k) = O\left(\frac{1}{k} S_0^{1.5} + S_0\right)$$

and when the number of robots is independent in the size of the region, the time complexity can be reduced to:

$$t_{success}(k) = O\left(S_0^{1.5}\right)$$

Namely:

Corollary 3

$$\left(\frac{8(|\partial F_0| - 1) \cdot (W(F_0) + k)}{k} + 2k < d\right) \Rightarrow \left(t_{success}(k) = O\left(S_0^{1.5}\right)\right)$$

Comparing this to the bound of Eq. 4, we see that the time complexity of the **SWEEP** protocol is much smaller than this of the nodes counting technique.

It should be mentioned though, that in [35] the authors clearly state that it is their belief that the coverage time of robots using nodes counting in grids is much smaller. This estimation is also demonstrated experimentally. However, it should also be pointed out that the methods described in [35] require the robots to have some sort of *marking* capability (namely, leaving trails along the region's tiles). Such a requirement is not necessary for the **SWEEP** protocol.

6 Regions with Obstacles

So far we have only dealt with simply connected regions, i.e. — regions with no "holes". In the case of a (connected) region with obstacles (i.e. holes) the simple **SWEEP**ing protocol will not work, due to the following "cycle" problem: eventually, each obstacle will be surrounded by critical points, and there will be a time when *all* boundary points of F will be critical (we shall call such a situation *useless*, as opposed to the *useful* state when some points are cleaned during a tour). As a cure to this problem, we suggest to add an "odoring" feature to our cleaners, that is — an agent will be able to label each tile it is going through by a small amount of "perfume" (this action may remind one of the pheromones left by ants during their walks). These labels will designate the directions according to which agents traveled through the labeled tiles. Upon getting to the useless state (detected by each agent due to no cleaning between two consecutive visits from the same entrance to the pivot point) an agent will continue to traverse the boundary, but will now look for a point which has a single label — that is, one that has odor pointing only on one direction of movement. If this tile does not have a neighbor which is entirely unlabeled, the agent will clean this point (despite its "criticality", as it is not really critical since it is necessarily part of a cycle around an obstacle) and then continue as in a useful state (see Fig. 14 for an example). This will open one cycle, hence, if there are m obstacles, we will need $\frac{m}{k}$ such tours before the region is completely clean. On the other hand, Lemma 3 no longer holds (see Fig. 16) since the boundary area can increase with time. However the boundary is always bounded above by the area. We now make the following conjecture:

Conjecture 1 Assume that k agents start at some boundary point of a non-simple connected region F_0 with s obstacles in it, and work according to the **SWEEP-WITH-OBSTACLES** protocol, and denote by $t_{success}(k)$ the time needed for this group to clean F_0. Then it holds that:

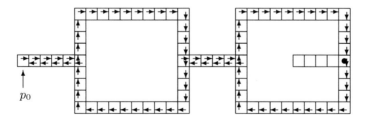

Fig. 14 An example of agents' odoring used in Sect. 6 for cleaning regions with obstacles. The figure demonstrates the labels on the region's tiles after an agent reached the *useless* state, in which it did not clean any tile in its last tour. Note that any of the tiles labeled with one direction only, can be cleaned while still preserving the region's connectivity. Note that from those tiles, only tiles with a neighbor which is not labeled with any direction can not be cleaned, since it is indeed a critical point. Such a point is marked in the figure by a *black circle*

$$\max\left\{2k,\left\lceil\frac{S_0}{k}\right\rceil,2l(F_0)\right\}\leq t_{success}(k)\leq 8\cdot S_0\left(\frac{W(F_0)}{k}+\left\lceil\frac{s}{k}\right\rceil\right)+2k \qquad (5)$$

where $S_0, l(F_0)$ and $W(F_0)$ denote the free area, length and width of F_0, respectively, and s is the number of obstacles in F_0.

Notice though, that the completion of the cleaning process is still guaranteed.

7 Experimental Results

A computer simulation, implementing the **SWEEP** protocol, was constructed. The protocol was run on several shapes of regions and for numbers of agents varying from 1 to 20. See Figs. 15 and 16 for some examples of the evolution of the layout with time. The gray level of each pixel designates the index of the agent that actually cleaned this point. The right side of Fig. 15 shows the same region and number of agents as in the left side of this Figure, but with randomly chosen initial locations at the corners. It can be seen that the dirty region is cleaned in a similar way. It should be said here that all the theory we developed in the previous sections (up to

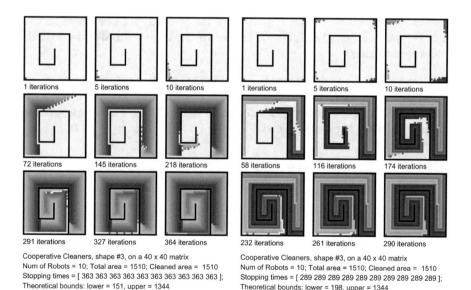

1 iterations   5 iterations   10 iterations   1 iterations   5 iterations   10 iterations

72 iterations   145 iterations   218 iterations   58 iterations   116 iterations   174 iterations

291 iterations   327 iterations   364 iterations   232 iterations   261 iterations   290 iterations

Cooperative Cleaners, shape #3, on a 40 x 40 matrix
Num of Robots = 10; Total area = 1510; Cleaned area = 1510
Stopping times = [363 363 363 363 363 363 363 363 363 363];
Theoretical bounds: lower = 151, upper = 1344

Cooperative Cleaners, shape #3, on a 40 x 40 matrix
Num of Robots = 10; Total area = 1510; Cleaned area = 1510
Stopping times = [289 289 289 289 289 289 289 289 289 289];
Theoretical bounds: lower = 198, upper = 1344

Fig. 15 Cooperative cleaners: maze, 10 agents (*left*), and 10 agents with random initial locations on the boundary (*right*)

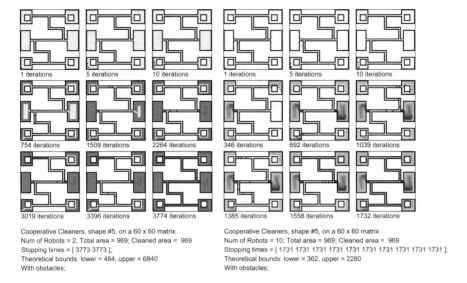

Cooperative Cleaners, shape #5, on a 60 x 60 matrix
Num of Robots = 2; Total area = 969; Cleaned area = 969
Stopping times = [3773 3773];
Theoretical bounds: lower = 484, upper = 6840
With obstacles;

Cooperative Cleaners, shape #5, on a 60 x 60 matrix
Num of Robots = 10; Total area = 969; Cleaned area = 969
Stopping times = [1731 1731 1731 1731 1731 1731 1731 1731 1731 1731];
Theoretical bounds: lower = 362, upper = 2280
With obstacles;

Fig. 16 Cooperative cleaners: 2 agents, 6 rooms with obstacles (*left*), 10 agents, 6 rooms with obstacles (*right*)

a small additive constant) applies to the case where the agents are initially located in randomly selected points on the boundary of F_0 (rather than starting from p_0). Figure 16 shows the evolution of the **SWEEP-WITH-OBSTACLES** protocol for the same regions with four additional obstacles in each, with 2 agents (left) and 10 agents (right).

Figure 17 depicts the timing results (as produced by computer simulations) describing the cleaning time as a function of the number of agents, compared to the theoretical bounds presented in previous sections (the cleaning time is normalized by the initial area of the dirty region). In Fig. 18 we show the results for the same figures with additional obstacles, together with the conjectured theoretical bounds. All the figures include their appropriate statistical values.

Additional result are presented in Fig. 19, comparing the various efficiency ratios obtained for initial dirty regions of various sizes. This is examined even further by Fig. 20, which shows the behavior of $\frac{t_{success}(50)}{t_{success}(1)}$ namely, the ratio between the cleaning time of 50 agents and the cleaning time required for a single agent. The change in this ratio can be seen, hinting the fact that as the initial region is larger, the addition of more agents can more efficiently decrease the cleaning time of the agents.

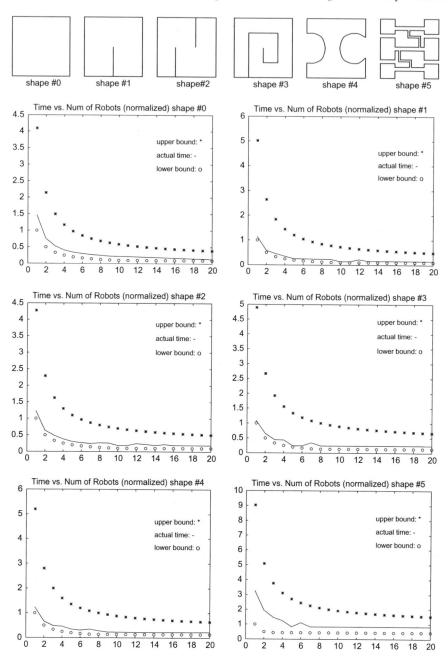

Fig. 17 Various shapes tested in **SWEEP** simulations and their normalized cleaning time $t(k) = \frac{t_{success}(k)}{S_0}$ as a function of the number of agents, k, for each region, together with the lower and upper bounds according to Theorem 2. For a confidence level of 95%, based on at least 10 simulation instances per each dirty region, the appropriate confidence interval is less than $\pm 6\%$

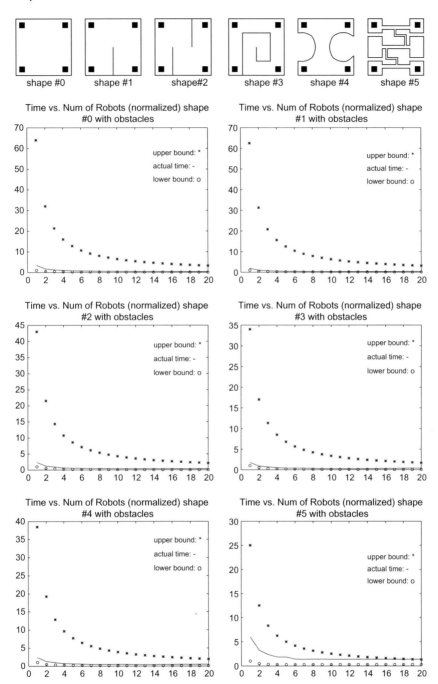

Fig. 18 Various shapes tested in **SWEEP-WITH-OBSTACLES** simulations and their normalized cleaning time $t(k) = \frac{t_{success}(k)}{S_0}$ as a function of the number of agents, k, together with the lower and upper bounds according to Theorem 2. For a confidence level of 95%, based on at least 10 simulation instances per each dirty region, the appropriate confidence interval is less than $\pm 10\%$

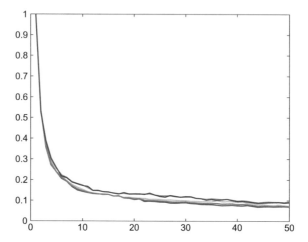

Fig. 19 Cooperative cleaners: from 1 to 50 agents, dirty regions of sizes 600 to 1500 tiles, 10 simulations per region's size, cleaning time normalized by the cleaning time of a single agent. Note that as the initial dirty area is larger, the ratio between $t_{success}(k)$ and $t_{success}(1)$ increases. In fact, comparing the time ratios of regions of sizes 600 and 1500 tiles, the difference reaches 30%. For a confidence level of 95% the confidence interval is less than $\pm 15\%$

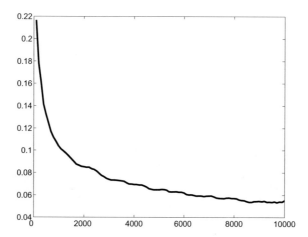

Fig. 20 Cooperative cleaners: graph demonstrates the behavior of $\frac{t_{success}(50)}{t_{success}(1)}$ for dirty regions of sizes 100 to 10000 tiles. 10 simulations were done per region's size. Note how as the initial dirty area gets larger $\frac{t_{success}(50)}{t_{success}(1)}$ decreases. The minimal value possible is of course $\frac{1}{50} = 0.02$. For a confidence level of 90% the confidence interval is less than $\pm 10\%$

8 Conclusion

Our cooperative **SWEEP**ing protocol can be considered as a case of social behavior in the sense of [34], where one induces multi-agent cooperation by forcing the agents to obey some simple "social" guidelines. This raises the question what happens if one agent malfunctions. We have shown that if less than k agents stop, the others will take over their responsibilities. But what if some agents start to cheat? Such adversaries will have catastrophic consequences, since a crazy agent may clean a critical point and disconnect the dirty region.

Another question of interest is the resolution of collisions between agents — the formation of clusters of agents located in the same tile, and delaying each other. In this work we resolve such a problem by artificially producing an implicitly agreed order among the agents, depending on their previous locations and movements. However, it is an interesting open question whether some random based method may achieve better results.

As can be seen throughout the analysis of our proposed collaborative protocol, the ability of a collaborative swarm of highly limited drones to efficiently scan a pre-defined region is strongly associated to the geometric features of this region. This topic was thoroughly analyzed in works such as [8, 11, 12], with a complexity result presented in [4, 5].

The cleaning problem discussed is also related to the geometric problem of pocket-machining, see [23] for details. An interesting problem of cleaning and maintenance of a system of pipes by an autonomous agent is discussed in [30]. The importance of cleaning hazardous waste by agents is described in [22].

Another question of interest is how to guarantee covering, at least with high probability, without using any external signs, using only the inter-agent collisions as an indicator for a good direction to proceed.

Considering the question of how fast can k agents clean a pre-defined contaminated region, it is of interest to notice here that an off-line version of the problem, that is: finding the shortest path that visits all grid points in F, where F is completely known in advance, is *NP-hard* even for a single agent. It is a corollary of the fact that Hamilton path in a non-simple grid-graph is *NP-complete* [25].

Finally, it is interesting to examine cases where the swarm have to operate not such under completion (e.g. ability to guarantee complete scan) or efficiency (e.g. scan time) constraints, but also under adversarial reactive behavior of the search region, namely – the ability of parts of the search space that were previously scanned to become "unscanned" after a certain portion of time. Such property of the problem is analyzed in the next chapters of this book, and was analyzed in works such as [6, 7, 10, 13, 32].

References

1. E.U. Acar, Y. Zhang, H. Choset, M. Schervish, A.G. Costa, R. Melamud, D.C. Lean, A. Gravelin, Path planning for robotic demining and development of a test platform, in *International Conference on Field and Service Robotics* (2001) pp. 161–168
2. M. Ahmadi, P. Stone, A multi-robot system for continuous area sweeping tasks, in *IEEE International Conference on Robotics and Automation, 2006 (ICRA 2006)* (2006)
3. Y. Altshuler, *Multi Agents Robotics in Dynamic Environments*. Ph.D. thesis, Israeli Institute of Technology, May 2010

4. Y. Altshuler, A.M. Bruckstein, The complexity of grid coverage by swarm robotics, in *ANTS 2010* (LNCS, 2010), pp. 536–543
5. Y. Altshuler, A.M. Bruckstein, Static and expanding grid coverage with ant robots: complexity results. Theoret. Comput. Sci. **412**(35), 4661–4674 (2011)
6. Y. Altshuler, A.M. Bruckstein, I.A. Wagner, Swarm robotics for a dynamic cleaning problem, in *IEEE Swarm Intelligence Symposium* (2005), pp. 209–216
7. Y. Altshuler, V. Yanovski, I.A. Wagner, A.M. Bruckstein, The cooperative hunters - efficient cooperative search for smart targets using uav swarms, in *Second International Conference on Informatics in Control, Automation and Robotics (ICINCO), the First International Workshop on Multi-Agent Robotic Systems (MARS)* (2005), pp. 165–170
8. Y. Altshuler, I.A. Wagner, A.M. Bruckstein, Shape factor's effect on a dynamic cleaners swarm, in *Third International Conference on Informatics in Control, Automation and Robotics (ICINCO), the Second International Workshop on Multi-Agent Robotic Systems (MARS)* (2006), pp. 13–21
9. Y. Altshuler, V. Yanovsky, I. Wagner, A. Bruckstein, Swarm intelligencesearchers, cleaners and hunters. *Swarm Intelligent Systems* (2006), pp. 93–132
10. Y. Altshuler, V. Yanovsky, A.M. Bruckstein, I.A. Wagner, Efficient cooperative search of smart targets using uav swarms. ROBOTICA **26**, 551–557 (2008)
11. Y. Altshuler, I.A. Wagner, A.M. Bruckstein, On swarm optimality in dynamic and symmetric environments. Economics **7**, 11 (2008)
12. Y. Altshuler, I.A. Wagner, A.M. Bruckstein, Collaborative exploration in grid domains, in *Sixth International Conference on Informatics in Control, Automation and Robotics (ICINCO)* (2009)
13. Y. Altshuler, I.A. Wagner, V. Yanovski, A.M. Bruckstein, Multi-agent cooperative cleaning of expanding domains. Int. J. Robot. Res. **30**, 1037–1071 (2010)
14. Y. Altshuler, A. Pentland, S. Bekhor, Y. Shiftan, A. Bruckstein, Optimal dynamic coverage infrastructure for large-scale fleets of reconnaissance uavs (2016), arXiv:1611.05735
15. R. Baeza-Yates, R. Schott, Parallel searching in the plane. Comput. Geom. **5**, 143–154 (1995)
16. M.A. Batalin, G.S. Sukhatme, Spreading out: a local approach to multi-robot coverage, in *6th International Symposium on Distributed Autonomous Robotics Systems* (2002)
17. V. Braitenberg, *Vehicles* (MIT Press, Cambridge, 1984)
18. Z. Butler, A. Rizzi, R. Hollis, Distributed coverage of rectilinear environments, in *Proceedings of the Workshop on the Algorithmic Foundations of Robotics* (2001)
19. D.C. Conner, A. Greenfield, P.N. Atkar, A.A. Rizzi, H. Choset, Paint deposition modeling for trajectory planning on automotive surfaces. IEEE Trans. Autom. Sci. Eng. **2**(4), 381–392 (2005)
20. M.P. Do-Carmo, *Differential Geometry of Curves and Surfaces* (Prentice-Hall, New-Jersey, 1976)
21. Y. Gabriely, E. Rimon, Competitive on-line coverage of grid environments by a mobile robot. Comput. Geom. **24**, 197–224 (2003)
22. S. Hedberg, Robots cleaning up hazardous waste, in *AI Expert* (1995), pp. 20–24
23. M. Held, On the computational geometry of pocket machining. *Lecture Notes in Computer Science* (1991)
24. D. Henrich, Space efficient region filling in raster graphics. Vis. Comput. **10**, 205–215 (1994)
25. A. Itai, C.H. Papadimitriou, J.L. Szwarefiter, Hamilton paths in grid graphs. SIAM J. Comput. **11**, 676–686 (1982)
26. S. Koenig, Y. Liu, Terrain coverage with ant robots: a simulation study, in *AGENTS'01* (2001)
27. S. Koenig, B. Szymanski, Y. Liu, Efficient and inefficient ant coverage methods. Ann. Math. Artif. Intell. **31**, 41–76 (2001)
28. D. Latimer, S. Srinivasa, V. Lee-Shue, S. Sonne, H. Choset, A. Hurst, Toward sensor based coverage with robot teams, in *Proceedings of the 2002 IEEE International Conference on Robotics and Automation* (2002)
29. T.W. Min, H.K. Yin, A decentralized approach for cooperative sweeping by multiple mobile robots, in *IEEE/RSJ International Conference on Intelligent Robots and Systems* (1998), pp. 380–385

30. W. Neubauer, Locomotion with articulated legs in pipes or ducts. Robot. Auton. Syst. **11**, 163–169 (1993)
31. M. Polycarpou, Y. Yang, K. Passino, A cooperative search framework for distributed agents, in *IEEE International Symposium on Intelligent Control* (2001), pp. 1–6
32. E. Regev, Y. Altshuler, A.M. Bruckstein, The cooperative cleaners problem in stochastic dynamic environments (2012), arXiv:1201.6322
33. I. Rekleitis, V. Lee-Shuey, A. Peng Newz, H. Choset, Limited communication, multi-robot team based coverage, in *IEEE International Conference on Robotics and Automation* (2004)
34. Y. Shoham, M. Tennenholtz, On social laws for artificial agent societies: off line design. AI J. **73**(1–2), 231–252 (1995)
35. J. Svennebring, S. Koenig, Building terrain-covering ant robots: a feasibility study. Auton. Robots **16**(3), 313–332 (2004)
36. I.A. Wagner, Y. Altshuler, V. Yanovski, A.M. Bruckstein, Cooperative cleaners: a study in ant robotics. Int. J. Robot. Res. (IJRR) **27**(1), 127–151 (2008)

Swarm Search of Expanding Regions in Grids: Lower Bounds

1 Introduction

One of the most interesting challenges for a swarm of robotic drones is the design and analysis of a multi-robotics system for searching areas (whose dimensions and shape are either known or unknown) [19, 25, 27, 30, 33, 34] or see [2] for a survey of search and evasion strategies. Interesting works to mention in this scope are those of [26, 35], where a swarm of ant-like robots is used for repeatedly covering an unknown area, using a real time search method called *node counting*. By using this method, the robots are shown to be able to efficiently perform such a coverage mission, and analytic bound for the coverage time are discussed.

While in most works the targets of the search mission were assumed to be idle, recent works considered dynamic targets, meaning — targets which after being detected by the searching robots, respond by performing various evasive maneuvers intended to prevent their interception.

In this chapter we discuss the problem concerning a given (yet potentially unknown) region that needs to be searched (or *"cleaned"*) by a swarm of autonomous robotic agents, namely – to have all of its 'tiles', or 'pixels' visited at least once by at least a single member of the swarm, and that this "cleaning" would be guaranteed to be completed within a finite (and as short as possible) time. We further assume that the region that needs to be searched *"expands"* over time, namely – that parts of it that were already "cleaned" (or searched) can become "contaminated" (or "unsearched") again. Under this assumption, the completion of the cleaning mission is achieved only when all of the region's pixels are in "clean" status (e.g. were visited by the drones, and did not become contaminated again yet). This dynamic variant of the problem assumes a deterministic evolution of the environment, simulating a spreading *contamination*, or *fire* [7].

This chapter is based on work previously published in parts in [3, 10, 14].

© Springer International Publishing AG 2018
Y. Altshuler et al., *Swarms and Network Intelligence in Search*,
Studies in Computational Intelligence 729, DOI 10.1007/978-3-319-63604-7_3

We assume that the swarm is collaborative, namely – that the various members of it are equipped with the same (and properly synchronized) software, and that each robot can acquire only the information which is available in its immediate vicinity, with the only way of inter-robot communication is by leaving traces on the common ground and sensing the traces left by other robots. Similar works can be found at [1, 22, 28, 29, 31].

Theoretical lower bounds for the problem are presented, as well as an impossibility result. A collaborative search protocol designed for a swarm of drones is presented, as well as a wealth of experimental results testing its performance and their comparison to the lower bounds. It should be stated that the paradigm concerning the drones used in this work is identical to this assumed for the static variant (as presented in the previous chapter, as well as in [37]). Namely — the drones used are assumed to be myopic, use no explicit long-range communication, and have only a small amount of memory (of size $O(\log k)$).

This chapter is organized as follows: the dynamic variant of the problem is formally defined in Sect. 2. The following analytic results are later discussed: Given a contaminated shape F_0 with an initial area S_0, and k agents employing *any* cleaning protocol, two lower bounds for S_t (the area of F at time t) are presented as Theorems 1 and 2 in Sect. 3. Using the above, an impossibility result is presented, as a constructive algorithm to generate contaminated regions which are impossible to clean, given any pair of (k and d) (discussed in Sect. 4). The cleaning protocol is presented in Sect. 5. Various experimental examples are presented in Sect. 6). We conclude this chapter with a detailed discussion concerning the isoperimetric inequality in grid domains, required by the proofs in Sect. 3, and a short discussion in Sect. 8. In addition, the chapter contains several terms, mechanisms and proofs that were discussed in previous chapters, but appear here for the purpose of self-containment (see Appendix).

2   The Dynamic "Cooperative Cleaners" Problem

The definition of the dynamic cooperative cleaners problem is very similar to that of the static variant, albeit with several required extensions. For the sake of readability, following is the complete self-contained definition of the problem.

We shall assume that the time is discrete.

Definition 1 Let the undirected graph $G(V, E)$ denote a two dimensional integer grid $\mathbf{Z}^2$, whose vertices (or "*tiles*") have a binary property called '*contamination*'. Let $cont_t(v)$ denote the contamination state of the tile v at time t, taking either the value "*on*" (for "dirty" or "contaminated") or "*off*" (for "*clean*").

For two vertices $v, u \in V$, the edge (v, u) may belong to E at time t only if both of the following hold:

- v and u are 4 *Neighbors* in G.
- $cont_t(v) = cont_t(u) = on$.

This however is a necessary but not a sufficient condition.

The edges of E represent the connectivity of the contaminated region. At $t = 0$ all the contaminated tiles are connected, namely:

$$(v, u) \in E_0 \iff (v, u \text{ are } 4 \text{ Neighbors in } G) \wedge (cont_0(v) = cont_0(u) = on)$$

Edges may be added to E only as a result of a contamination spread and can be removed only while contaminated tiles are cleaned by the agents (see below).

Definition 2 Let $F_t(V_{F_t}, E_t)$ be the contaminated sub-graph of G at time t, i.e.:

$$V_{F_t} = \{v \in G \mid cont_t(v) = on\}$$

We assume that F_0 is a single simply-connected component (the actions of the agents will be so designed that this property will be preserved).

Let a group of k agents that can move on the grid G (moving from a tile to its neighbor in one time step) be placed at time t_0 on F_0, at point $p_0 \in V_{F_t}$.

Each agent is equipped with a sensor capable of telling the *contamination* status of all tiles in the digital sphere of diameter 7, which surrounds the agent. An agent is also aware of other agents which are located in these tiles, and all the agents agree on a common direction. Each tile may contain any number of agents simultaneously. This information will later be required by the agents' cleaning protocol. Each agent is equipped with a memory of size $O(\log k)$ bits (used later as an agents counter).

When an agent moves to a tile v, it has the possibility of cleaning this tile (i.e. causing $cont(v)$ to become *off*). Once an agent cleaned a tile v, all the edges touching v are removed, namely:

$$E_{t+1} = E_t \setminus \{(v, u) \mid (v, u) \in E_t \wedge cont(v) = off\}$$

The agents do not have any prior knowledge of the shape or size of the sub-graph F_0 except that it is a single and simply connected component.

The contaminated region F_t is assumed to be surrounded at its boundary by a "rubber-like" elastic barrier, dynamically reshaping itself to fit the evolution of the contaminated region over time (the barrier is derived from the edges of E_t, as demonstrated in Fig. 1). This process can be thought of as a water filed rubber balloon — while the shape of the water contained in it may change, as well as its volume the balloon keeps bounding it tightly. This barrier is intended to guarantee the preservation of the simple connectivity of F_t, in order to allow the agents to perform their cleaning activities. The need for such a barrier is discussed in Sect. 8.

When an agent cleans a contaminated tile, the barrier is withdrawn, in order to free the void previously occupied by the cleaned tile. This is demonstrated in Fig. 2.

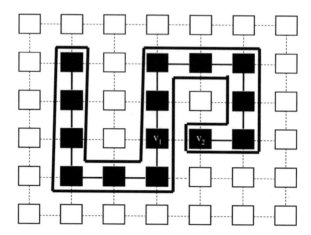

Fig. 1 An illustration of the barrier surrounding the contaminated region, derived from the edges of E_t. Notice that when an edge connecting two contaminated tiles are not in E_t, those tiles are "on the other side" of the barrier. For example, observe the vertices v_1 and v_2, which are 4 *Neighbors* in F_t (namely, the Manhattan distance between them in G equals one), however — as $\neg((v_1, v_2) \in E_t)$ the geodesic distance between them is significantly greater

Fig. 2 An illustration of the evolution of the elastic barrier as a result of the movement and cleaning of an agent according to the **SWEEP** protocol, (as described in Figs. 25 and 26)

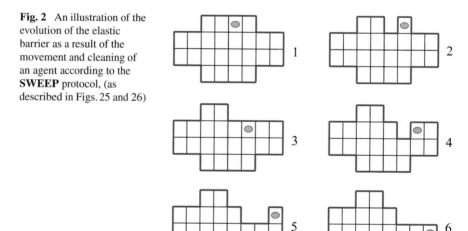

Definition 3 Let d denote the number of time steps between contamination spreads. That is, if $t = nd$ for some positive integer n, then:

$$\forall v \in F_t \; \forall u \in 4 \, Neighbors(v) \, , \; cont_{t+1}(u) = on$$

While the contamination spreads, the elastic barrier stretches accordingly. First, all the clean tiles located to the right of a contaminated tile are becoming contaminated. Then, the clean tiles located below contaminated tiles, followed by clean tiles located to the left of such tiles. Finally, clean tiles located above a contaminated tile are affected. The barrier itself it implicitly derived from the edges connecting the tiles. Namely, it is the boundary which surrounds the connected contaminated tiles. This process is illustrated in Figs. 3 and 4, and can be formalized as follows:

Definition 4 Elastic barrier's expansion as a result of a contamination spread:

1. Let $\widehat{V}_t = \emptyset$ be a set of vertices and let $\widehat{E}_t = \emptyset$ be a set of edges
2. For vertex u located at (x_u, y_u) let:

   - $Neighbor_{right}(u)$ denote the vertex located at $(x_u + 1, y_u)$
   - $Neighbor_{left}(u)$ denote the vertex located at $(x_u - 1, y_u)$
   - $Neighbor_{up}(u)$ denote the vertex located at $(x_u, y_u + 1)$
   - $Neighbor_{down}(u)$ denote the vertex located at $(x_u + 1, y_u - 1)$

3. Let $d_G(v, u)$ denote the number of edges between v and u in the graph G
4. Add new vertices to the right —

   (a) $V_{temp} = \left\{ v \mid \neg(v \in V_{F_t} \cup \widehat{V}_t) \wedge \left(\exists u \in V_{F_t}, \ v = Neighbor_{right}(u) \right) \right\}$
   (b) $\widehat{E}_t \leftarrow \widehat{E}_t \cup \left\{ (v, u) \mid (v \in V_{temp}) \wedge (v = Neighbor_{right}(u)) \right\}$
   (c) $\widehat{E}_t \leftarrow \widehat{E}_t \cup \left\{ (v, u) \ \middle| \ \begin{array}{l} (v \in \widehat{V}_t) \wedge (v = Neighbor_{right}(u)) \wedge \\ (u \in V_{F_t}) \wedge (d_{G(V_{F_t} \cup \widehat{V}_t, E_t \cup \widehat{E}t)}(v, u) \le 3) \end{array} \right\}$
   (d) $\widehat{V}_t \leftarrow \widehat{V}_t \cup V_{temp}$

5. Add new vertices to the bottom —

   (a) $V_{temp} = \left\{ v \mid \neg(v \in V_{F_t} \cup \widehat{V}_t) \wedge \left(\exists u \in V_{F_t}, \ v = Neighbor_{down}(u) \right) \right\}$
   (b) $\widehat{E}_t \leftarrow \widehat{E}_t \cup \left\{ (v, u) \mid (v \in V_{temp}) \wedge (v = Neighbor_{down}(u)) \right\}$
   (c) $\widehat{E}_t \leftarrow \widehat{E}_t \cup \left\{ (v, u) \ \middle| \ \begin{array}{l} (v \in \widehat{V}_t) \wedge (v = Neighbor_{down}(u)) \wedge \\ (u \in V_{F_t}) \wedge (d_{G(V_{F_t} \cup \widehat{V}_t, E_t \cup \widehat{E}t)}(v, u) \le 3) \end{array} \right\}$
   (d) $\widehat{V}_t \leftarrow \widehat{V}_t \cup V_{temp}$

6. Add new vertices to the left —

   (a) $V_{temp} = \left\{ v \mid \neg(v \in V_{F_t} \cup \widehat{V}_t) \wedge \left(\exists u \in V_{F_t}, \ v = Neighbor_{left}(u) \right) \right\}$
   (b) $\widehat{E}_t \leftarrow \widehat{E}_t \cup \left\{ (v, u) \mid (v \in V_{temp}) \wedge (v = Neighbor_{leftt}(u)) \right\}$
   (c) $\widehat{E}_t \leftarrow \widehat{E}_t \cup \left\{ (v, u) \ \middle| \ \begin{array}{l} (v \in \widehat{V}_t) \wedge (v = Neighbor_{left}(u)) \wedge \\ (u \in V_{F_t}) \wedge (d_{G(V_{F_t} \cup \widehat{V}_t, E_t \cup \widehat{E}t)}(v, u) \le 3) \end{array} \right\}$
   (d) $\widehat{V}_t \leftarrow \widehat{V}_t \cup V_{temp}$

7. Add new vertices to the top —

   (a) $V_{temp} = \left\{ v \mid \neg(v \in V_{F_t} \cup \widehat{V}_t) \wedge \left(\exists u \in V_{F_t}, \ v = Neighbor_{up}(u) \right) \right\}$
   (b) $\widehat{E}_t \leftarrow \widehat{E}_t \cup \left\{ (v, u) \mid (v \in V_{temp}) \wedge (v = Neighbor_{up}(u)) \right\}$
   (c) $\widehat{E}_t \leftarrow \widehat{E}_t \cup \left\{ (v, u) \ \middle| \ \begin{array}{l} (v \in \widehat{V}_t) \wedge (v = Neighbor_{up}(u)) \wedge \\ (u \in V_{F_t}) \wedge (d_{G(V_{F_t} \cup \widehat{V}_t, E_t \cup \widehat{E}t)}(v, u) \le 3) \end{array} \right\}$
   (d) $\widehat{V}_t \leftarrow \widehat{V}_t \cup V_{temp}$

8. $V_{F_t} \leftarrow V_{F_t} \cup \widehat{V}_t$

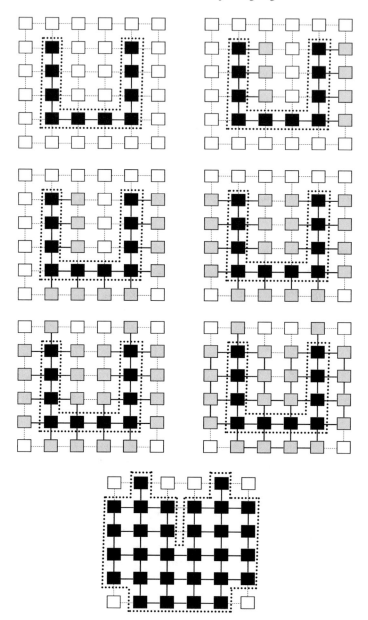

Fig. 3 An illustration of the barrier expansion process as a result of a contamination spread. The example starts from the top left chart, going right

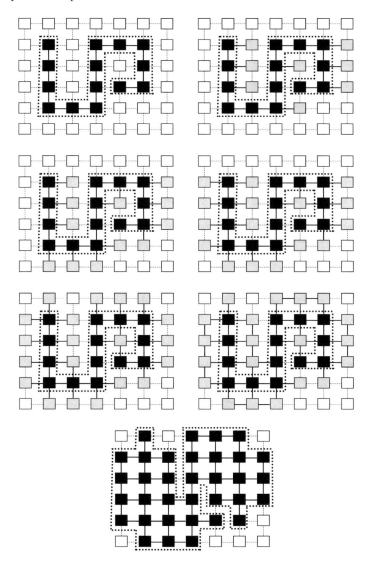

Fig. 4 An illustration of the barrier expansion process as a result of a contamination spread. The example starts from the top left chart, going right

9. $E_t \leftarrow E_t \cup \widehat{E}_t$
10. $E_t \leftarrow E_t \cup \big\{(v, u) \mid (v, u \in V_{F_t}) \wedge (v = 4 \, Neighbors(u)) \wedge (d_{F_t}(v, u) \leq 3)\big\}.$

For the agents who travel along the tiles of F, the barrier signals the boundary of the contaminated region. When an agent detects a contaminated tile which is "on the other side" of the barrier it treats this tile as though it was a clean tile (this aspect is important to remember when reviewing the agents' behavior while following the

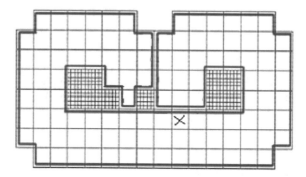

Fig. 5 Observe how the stretching of the barrier is generating "double-fronts" after a contamination spread. Note that had the contaminated parts been allowed to merge, the simple connectivity of the region would have not been kept. In addition, observe how an agent located in the tile marked by an "X" is still kept on the boundary of the contaminated region, due to this mechanism

cleaning protocol). For example, examining the region which appears in Fig. 5 and assuming an agent located in the "X" marked tile, then while looking "upwards", this agent acts as though the tile above it is clean (as the contaminated tiles are located behind the barrier and are thus masked as clean ones).

As in the case of the static variant, it is the agents' goal to clean G by eliminating the contamination entirely, meaning that the agents must ensure that:

$$\exists t_{success} \ s.t \ F_{t_{success}} = \emptyset$$

In addition, it is desired that time $t_{success}$ will be minimal.

In addition, and as stated previously, we still impose the restriction of no central control and a fully 'de-centralization' system.

3   A Protocol-Agnostic Lower Bound for the Cleaning Time

Due to the dynamic nature of the problem, the shape of the contaminated region can change dramatically during the cleaning process. Therefore, since we know no easy way to decide whether k agents can successfully clean an instance of the *Dynamic Cooperative Cleaners* problem, producing bounds for the efficiency of any proposed cleaning protocol is important for estimating its efficiency.

Let S_t denote the size of the contaminated region F at time t, namely the number of tiles (i.e. grid points) in F.

Let us now discuss a general lower bound for the problem of dynamic cleaning. The bound is general in the sense that it bounds the problem, not a specific cleaning protocol for it. Therefore, it is applicable for any cleaning protocol which might be used by the agents. This is achieved by showing that at a specific time t, $S_t > 0$, namely — the mission could not be completed until that time (regardless of the cleaning protocol used), as the completion of the cleaning mission at time t means that $S_t = 0$.

Theorem 1 *Using any cleaning protocol, the area of the contaminated region at time t can be recursively lower bounded, as follows:*

$$S_{t+d} \geq S_t - d \cdot k + \left\lfloor 2\sqrt{2 \cdot (S_t - d \cdot k)} - 1 \right\rfloor$$

Proof Notice that a lower bound for the cleaning time can be obtained by assuming that the agents are working with maximal efficiency, meaning that each time step every agent cleans exactly one tile. After $(d - 1)$ time steps k agents will therefore clean $k \cdot (d - 1)$ tiles, and thus we know that:

$$S_{d-1} \geq S_0 - (d - 1) \cdot k$$

In the d-th time step, the agents clean another portion of k tiles, but the remaining contaminated tiles spread their contamination to their 4 *Neighbors* and cause new tiles to become contaminated. We are interested in the *minimal* number of tiles which can become contaminated at this stage. The minimal number of 4 *Neighbors* of any number of tiles is achieved when the tiles are organized in the shape of a "digital sphere" (see a complete proof in Sect. 7), as demonstrated in Fig. 6. Therefore, for a region of any given area S the minimal number of its 4 *Neighbors* is at least $2\sqrt{2 \cdot S_t} - 1$. As the number of tiles must be an integer value, we use $\left\lfloor 2\sqrt{2 \cdot S_t} - 1 \right\rfloor$ to remain on the safe side.

After the d-th time step we thus have the following:

$$S_d \geq S_0 - d \cdot k + \left\lfloor 2\sqrt{2 \cdot (S_0 - d \cdot k)} - 1 \right\rfloor$$

and the rest is implied.

An illustration of Theorem 1 is presented is Fig. 7.

The following result is a direct lower bound for the size of the contaminated region, based on the recursive bound of Theorem 1.

Fig. 6 The *left chart* shows an example of a 'sphere' in the grid. Notice that this sphere has only 16 tiles in its 4 *Neighbors* while the right chart, containing a shape of the same area, has 20 tiles in its 4 *Neighbors*

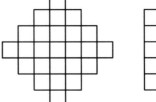

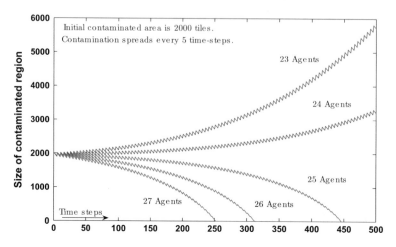

Fig. 7 Theorem 1 illustrated. The graph depicts a lower bound over the size of the contaminated region as a function of t, given various number of cleaning agents (while the initial area of the contaminated region is $S_0 = 2000$ and the contamination spreading latency is $d = 5$). The lower bound over the cleaning time can be interpreted as stating that the time for completion of the cleaning job will be higher than the point at which the graph of S_t hits 0. For example, observe how a complete cleaning is shown to be impossible for any swarm with 24 agents or less. Naturally, as Theorem 1 only takes into account the initial size of the contaminated region, different region's size evolutions are expected to be seen for regions of various shapes, but of the same initial size — see such examples in Sect. 6. Note that as this bound is generic, hence independent of the cleaning protocol used by the agents, the actual cleaning protocol used will have a significant effect on the actual cleaning efficiency

Theorem 2 *Using any cleaning protocol, a lower bound for the area of the contaminated region at time $t = i \cdot d$ for some $i \in \mathbb{N}$ is the minimal positive S_t, which is a solution of the following equation:*

$$2i = \sqrt{2(S_t - dk) - 1} - \sqrt{2(S_0 - dk) - 1} + \ln\left(\frac{\sqrt{2(S_t - dk) - 1} - \frac{dk}{2}}{\sqrt{2(S_0 - dk) - 1} - \frac{dk}{2}}\right)^{\frac{dk}{2}}$$

Proof Observe that by denoting $y_i \triangleq S_{id}$ Theorem 1 can be written as:

$$y_{i+1} - y_i \geq \left\lfloor 2\sqrt{2 \cdot (y_i - d \cdot k) - 1} \right\rfloor - d \cdot k$$

Searching for the minimal area we can look at the equation:

$$y_{i+1} - y_i = \left\lfloor 2\sqrt{2 \cdot (y_i - d \cdot k) - 1} \right\rfloor - d \cdot k$$

By dividing both sides by $\Delta i = 1$ we obtain:

$$y_{i+1} - y_i \triangleq y' = \left\lfloor \sqrt{8y - 8\left(d \cdot k + \frac{1}{2}\right)} \right\rfloor - d \cdot k \qquad (1)$$

Notice that the values of y' might be positive (stating an increase in the area), negative (stating a decrease in the area), or complex numbers (stating that the area is smaller than $d \cdot k$, and will therefore be cleaned before the next spread).

Let us denote $x^2 \triangleq 8y - 8\left(d \cdot k + \frac{1}{2}\right)$. After calculating the derivative of both sides of this expression we see that:

$$2x \cdot x' = 8y'$$

and after using the definition of y' of Eq. 1 we see that:

$$2x \cdot \frac{dx}{di} = 2x \cdot x' = 8\left\lfloor \sqrt{8y - 8\left(d \cdot k + \frac{1}{2}\right)} \right\rfloor - 8d \cdot k \leq 8\,(x - d \cdot k) \qquad (2)$$

From Eq. 2 a definition of di can be extracted:

$$di \geq \frac{1}{8} \cdot \frac{2x}{x - d \cdot k} dx \geq \frac{1}{8} \cdot \frac{2x - 2d \cdot k + 2d \cdot k}{x - d \cdot k} dx \geq \frac{1}{4}\left(1 + \frac{d \cdot k}{x - d \cdot k}\right) dx$$

The value of x can be achieved by integrating the previous expression as follows (notice that we are interested in the equality of the two expressions):

$$\int_{i_0}^{i} di = \int_{x_0}^{x} \frac{1}{4}\left(1 + \frac{d \cdot k}{x - d \cdot k}\right) dx$$

After the integration we can see that:

$$i \Big|_{i_0}^{i} = \frac{1}{4}\left(x + d \cdot k \ln\left(x - d \cdot k\right)\right) \Big|_{x_0}^{x}$$

and after assigning $i_0 = 0$:

$$4i = x - x_0 + d \cdot k \ln \frac{x - d \cdot k}{x_0 - d \cdot k}$$

Returning back to y we get:

$$2i = \sqrt{2(y - d \cdot k) - 1} - \sqrt{2(y_0 - d \cdot k) - 1} + \ln\left(\frac{\sqrt{2(y - d \cdot k) - 1} - \frac{d \cdot k}{2}}{\sqrt{2(y_0 - d \cdot k) - 1} - \frac{d \cdot k}{2}}\right)^{\frac{d \cdot k}{2}}$$

Fig. 8 Theorem 2: These graphs depict lower bounds for the cleaning time of several contaminated regions (the Y axis), as a function of the number of agents (the X axis, starting from the lowest number of agents which produces a successful cleaning). A comparison to the naive $f(k) = \frac{S_0}{k}$ function, which holds for static environments, is also included (the *lower curve*, marked in *red*). Notice again that this bound is generic, and therefore will hold for any cleaning protocol used

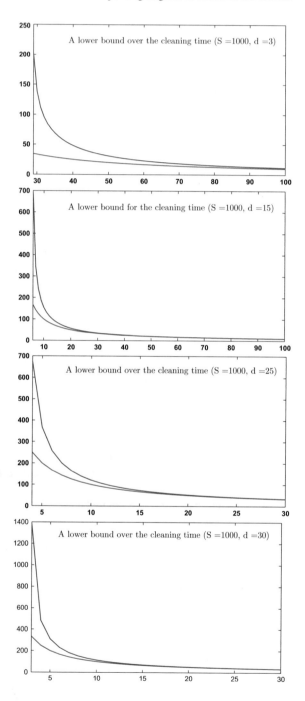

Fig. 9 Theorem 2: These graphs depict lower bounds for the cleaning time of several contaminated regions (the Y axis), as a function of the number of agents (the X axis), for various values of d. Notice again that this bound is generic, and therefore will hold for any cleaning protocol used

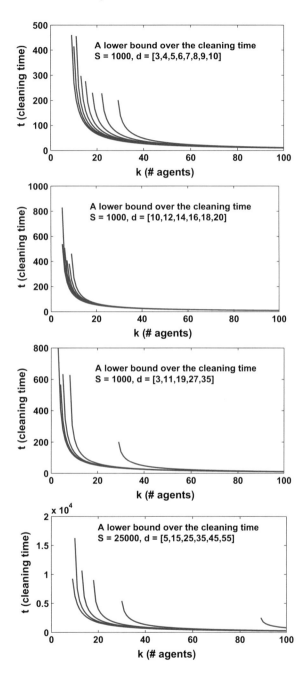

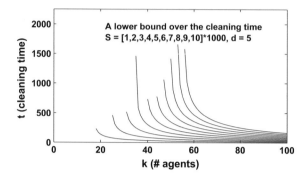

Fig. 10 Theorem 2: These graphs depict lower bounds for the cleaning time of several contaminated regions (the Y axis), as a function of the number of agents (the X axis), for various values of the initial area S_0

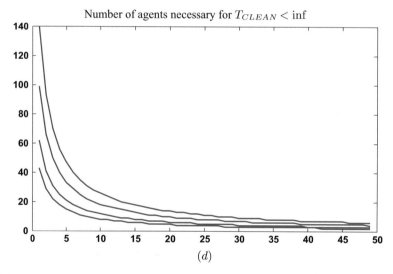

Fig. 11 Theorem 2: This graph represents lower bounds for the minimal number of agents required to guarantee a successful cleaning of several contaminated regions, as a function of the contamination spreading speed d. The *lower curve* depicts the minimal number of agents for an initial region of $S_0 = 1000$, for spreadings speeds between 2 and 50. Similarly, the other curves represent a similar functions for $S_0 = 2000, 5000, 10000$, respectively

and returning to the original size variable S_t we see that:

$$2i = \sqrt{2(S_{id} - dk) - 1} - \sqrt{2(S_0 - dk) - 1} + \ln\left(\frac{\sqrt{2(S_{id} - dk) - 1} - \frac{dk}{2}}{\sqrt{2(S_0 - dk) - 1} - \frac{dk}{2}}\right)^{\frac{dk}{2}}$$

and the rest of the Theorem is implied.

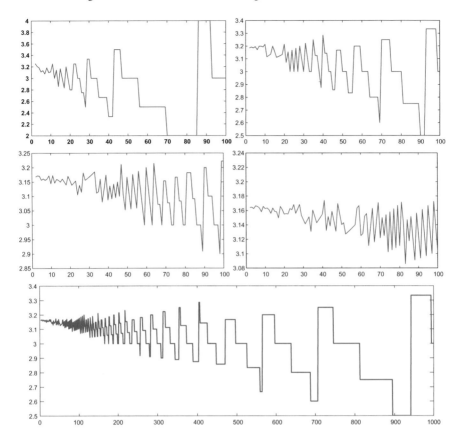

Fig. 12 An example of Theorem 2. These graphs demonstrate the influence of the change in the area of the initial contaminated region over the change in the lower bound for the minimal number of agents required to guarantee a successful cleaning of the region, namely — the ratio $\frac{k(S_0,d)}{k(\frac{1}{10}S_0,d)}$ (as a function of d), when $k(S,d)$ represents the lower bound over the minimal number of agents, for a region of area S and a spreading speed d. The values of S_0 of the *four upper graphs* are 10000, 100000, 1000000 and 10000000, while the values of d are between 2 and 100. A continuation of the last graph (i.e. in which $S_0 = 10000000$) appears in the *lower graph*, in which $2 \leq d \leq 1000$. Observing these graphs it can be seen that, as expected, an increase in d results in a general decrease in the agents' ratio (disregarding the oscillatory nature of the function). The reason is that when d is increased, the problem becomes "easier", and therefore the number of agents required to complete the cleaning is smaller (regardless of the cleaning time itself). As to the effect of the initial area of the region — the ratio between the agents required for cleaning an area of S_0 tiles and the agents required for cleaning an area of $\frac{1}{10} S_0$ tiles is generally constant for smaller values of d (and surrounds $\sqrt{10}$), regardless of the ratio between S_0 and $\frac{1}{10}$

Theorem 2 provides an easy way of calculating the minimal possible time it takes a given number of agents to clean some portion of any given contaminated region (and specifically — to clean it entirely). A lower bound for the cleaning time of k agents for various contaminated regions based on this Theorem appears in Figs. 8, 9 and 10.

Fig. 13 An example of
Theorem 2, illustrating the
influence of the change in the
contaminated region's initial
size over the lower bound of
the cleaning time of the
smallest group of agents
which is capable of cleaning
it. Namely, the graphs
demonstrate
$R_{S_0}(\alpha) \triangleq \frac{T_{min}(\alpha \cdot S_0, D)}{T_{min}(S_0, D)}$ as a
function of α. Notice that the
value of $R_{S_0}(\alpha)$ is agnostic
to the selection of the initial
S_0 or of D

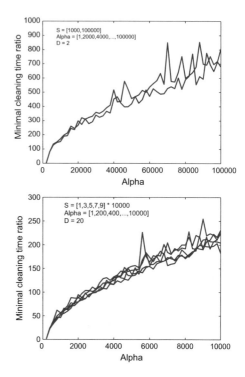

A lower bound for the minimal number of agents required to guarantee a successful cleaning of a contaminated region, as a function of the contamination spreading speed, is demonstrated in Fig. 11, for several values of the initial contaminated area. The influence of the initial contaminated area over the minimal number of agents required to guarantee a complete cleaning is demonstrated in Fig. 12, in which the ratio $\frac{K_{min}(S_0, d)}{K_{min}(\frac{1}{10}S_0, d)}$ (as a function of d) is presented (suggesting that when $K_{min}(n, d)$ denotes the minimal number of agents required to enable a cleaning of a contaminated region of initial contaminated region of n tiles, for some contamination spreading d, then $K_{min}(\alpha \cdot n, d) \approx \sqrt{\alpha} \cdot K_{min}(n, d)$).

Another interesting illustration of Theorem 2 is presented in Fig. 13, where the ratio $R_{S_0}(\alpha) \triangleq \frac{T_{min}(\alpha \cdot S_0, D)}{T_{min}(S_0, D)}$ (where $T_{min}(n, D)$ denotes the average lower bound for the cleaning time of $K_{min}(n, d)$ agents, where $d = [2, 3, 4, \ldots, D]$) is calculated for various values of S_0 (and for $D = 20$). Based on these examples it can be conjectured that $R_{S_0}(\alpha) \approx 2\sqrt{\alpha}$ regardless of S_0 and D,

4 Impossibility Result

While designing a system intended to use cleaning agents in order to guarantee a successful completion of the cooperative cleaners problem, various cleaning methods can be used. This, combined with the dramatic changes the initial geometric properties of the contaminated region can undergo, yields a great uncertainty as to the cleaning time of a group of agents using any protocol (and to that matter, as to the possibility of guaranteeing such a successful completion, to begin with). While the theoretical lower bound shown in Sect. 3 as well as the upper bounds which are presented in the second part of this work can assist in decreasing this uncertainty, one might alternatively be concerned with the opposite question, namely — how can we guarantee that a group of k agents *will not* be able to clean a contaminated region entirely (regardless of the cleaning protocol which is being used, or the agents' sensors or communication capabilities). One solution to this problem is offered by Theorem 3, as follows:

Theorem 3 *Using any cleaning protocol, k agents cleaning a contaminated region of size S_0, which spreads every d time steps, will not be able to clean it if:*

$$S_0 > \left\lceil \frac{1}{8}d^2k^2 + dk + \frac{1}{2} \right\rceil$$

Proof We shall require that the size of the region increases between each two spreads (thus generating an ever expanding region, impossible to clean), or in other words, we require that:

$$S_d - S_0 > 0$$

Using Theorem 1 we know that regardless of the specific protocol used:

$$S_d \geq S_0 - d \cdot k + 2\sqrt{2 \cdot (S_0 - d \cdot k) - 1}$$

and therefore:

$$2\sqrt{2 \cdot (S_0 - d \cdot k) - 1} > d \cdot k$$

After some arithmetics we see that:

$$S_0 > \left\lceil \frac{1}{8}d^2k^2 + dk + \frac{1}{2} \right\rceil.$$

See a demonstration of Theorem 3 in Fig. 14.

5   The Cleaning Protocol

In order to solve the *Dynamic Cooperative Cleaners* problem we use the same cleaning protocol which was proposed for the static variant of the problem, as presented in the previous chapter, namely – the **SWEEP** protocol (see the complete definition of the protocol either in the previous chapter, or in Appendix). However, there are several implications to the fact that the contamination is now assumed to be expanding over time. First, the analysis of the protocol's performance becomes somewhat more complicated (moreover, the protocol's correctness itself must be redemonstrated). Second, a group of k agents can no longer be assumed to clean a contaminated region given enough time (as shown for example in Sect. 4). Rather, the group of agents must be large enough, in order to guarantee a successful completion of the task. Finally, agents acting according to the **SWEEP** protocol, are expected to face a scenario which did not take place when the contaminated region was assumed to be static, namely — a spread of the contamination. Specifically, upon contamination spread, agents might find themselves located in tiles which are no longer boundary tiles. Once this happens, an agent will move towards the boundary tile which implements the continuation of his planned traversal along the region (this is executed through the movement to the *rightmost* neighbor, as defined in Definition 11).

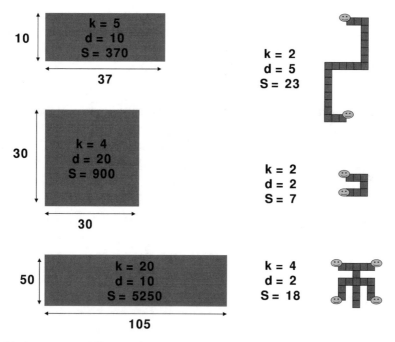

Fig. 14 An example of Theorem 3, presenting several examples for initial regions, spreading latencies, and number of agents, which are guaranteed to be impossible for cleaning

Additional technical mechanisms which were discussed in the previous chapter (such as the *Pivot Point*, *Signaling*, *Connectivity Preservation*, *Agents' Synchronization*, *Clustering Problem* and *Mission Termination*) remain unaffected by the dynamic nature of the problem. For the purpose of making this chapter self-contained, they are redefined in Appendix.

6 Experimental Results

A computer simulation, implementing the **SWEEP** cleaning protocol, was constructed (this simulated environment is currently available online at the Intelligent Systems Laboratory's website, at the Computer Science Department of the Technion). Exhaustive simulations were carried out, examining the cleaning activity of the protocol for regions of various geometric features.

Figures 15, 16 and 17 depict the cleaning process of several contaminated regions — for each one, the cleaning time for various number of agents is presented. As the agents' cleaning time is influenced by their initial location, for every number of agents, a multitude of simulations were performed, one for each possible location (namely, the total number of simulations for any number of agents equaled $|\partial F_0|$). The maximal and minimal cleaning times for each number of agents is presented, as well as the average cleaning time (calculated for all the initial locations). Figure 18 demonstrates how placing the agents at different starting points allows them to successfully clean regions which were impossible to clean when the agents were concentrated at a single starting point. Figure 19 demonstrates the dynamic nature of the agents' cleaning efficiency, by comparing the contaminated region's size to the whereabouts of the agents at various times. Figure 20 demonstrates the changes in the geometric features of a region while it is being cleaned, as well as the influence the number of agents has on the efficiency of the cleaning.

7 Isoperimetric Inequality on the Grid

The results that are discussed in this section are also thoroughly analyzed in [4, 11, 36].

Theorem 4 *For all the shapes of area S, let F be the shape with the minimal number of clean 4 Neighbors, then the number of clean 4 Neighbors of F is at least the number of 4 Neighbors of the largest digital sphere of size at most S.*

Proof

Definition 5 Let l_{ru} and l_{ld} denote two infinite sets of tiles, each organized as a straight line slanted by 45 degrees (in lower right direction). Similarly, let l_{lu} and l_{rd} denote two infinite sets of tiles, each organized as a straight line slanted by 45 degrees (in lower left direction).

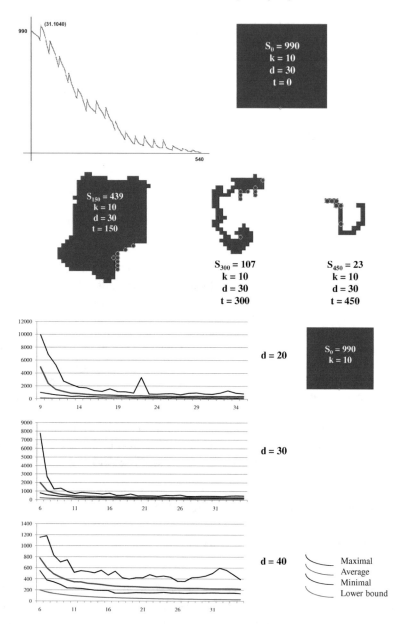

Fig. 15 An example of a cleaning mission by agents using the **SWEEP** cleaning protocol, trying to clean a contaminated square (see the original contaminated shape at the *top, on the right*). A graph of the size of the region as a function of the time, for 10 agents and spreading speed of $d = 30$ appears on the *left top corner*, as well as size and shape of the region at several times during the cleaning. The three graphs at the bottom depict the cleaning time (maximal, minimal and average) for various numbers of agents, and for three values of spreading speeds. In addition, the cleaning times are compared to the lower bound, as appears in Theorem 2

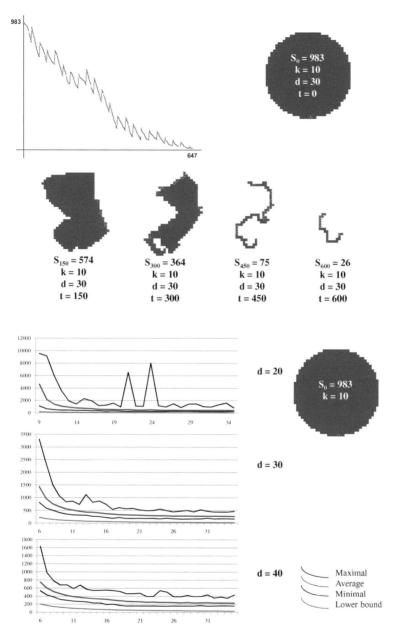

Fig. 16 An example of a cleaning mission by agents using the **SWEEP** cleaning protocol, trying to clean a contaminated circle (see the original contaminated shape at the *top, on the right*). A graph of the size of the region as a function of the time, for 10 agents and spreading speed of $d = 30$ appears on the *left top corner*, as well as size and shape of the region at several times during the cleaning. The three graphs at the bottom depict the cleaning time (maximal, minimal and average) for various numbers of agents, and for three values of spreading speeds. In addition, the cleaning times are compared to the lower bound, as appears in Theorem 2

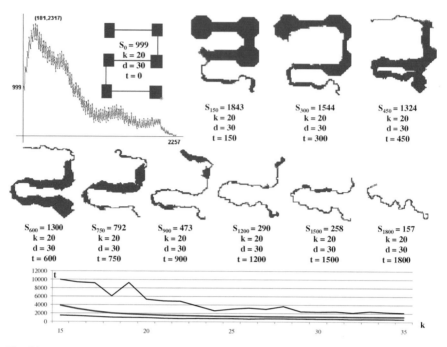

Fig. 17 An example of a cleaning mission by agents using the **SWEEP** cleaning protocol, trying to clean a contaminated region in the shape of several rooms, connected by narrow corridors (see the original contaminated shape at the *top, on the right*). A graph of the size of the region as a function of the time, for 20 agents and spreading speed of $d = 30$ appears on the *left top corner*, as well as size and shape of the region at several times during the cleaning. The graph at the bottom depicts the cleaning time (maximal, minimal and average) for various numbers of agents, and for a spreading speed of $d = 30$

For any shape F, let l_{ru} be place above and to the right of F, let l_{lu} be place above and to the left of F, let l_{rd} be place below and to the right of F and let l_{ld} be place below and to the left of F. Let us move l_{ru} and l_{rd} to the left, until they touch F. In addition, let us move l_{lu} and l_{ld} to the right, until they touch F. Note that the four lines form a bounding slanted rectangle around F.

Definition 6 Let us remove all the tiles of l_{ru}, l_{lu}, l_{rd} and l_{ld} which are not part of this rectangle, and denote the remaining tiles by *slanted-bounding-rectangle(F)*.

Definition 7 For each of the tiles sets l_{ru}, l_{rd}, l_{ld} and l_{lu} let us denote the last tiles of F that are 4 *Neighbors* of the sets (assuming clockwise movement) by 1, 2, 3 and 4 respectively.

An example appears in Fig. 21.

Let us project all the tiles of *slanted-bounding-rectangle(F)* between points 1 and 2 left, the points between 2 and 3 upwards, the points between 3 and 4 right and the points between 4 and 1 downwards, until they become 4 *Neighbors* of the tiles of F.

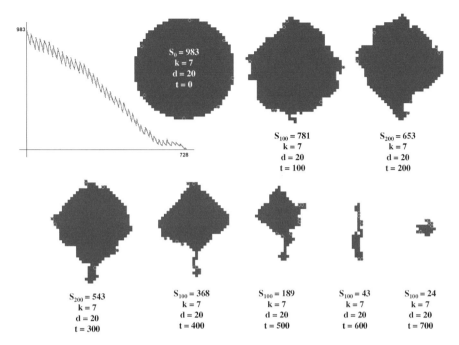

Fig. 18 The graph presents the cleaning process of a (relatively large) digital sphere. Observe how the size of the region increases at various times. Examining the appearance of the region at those times reveals that the majority of agents are being located at the vicinity of the initial location of the agents (a point which is artificially kept critical, according to the **SWEEP** protocol) — see snapshots at the bottom. After several traversals of the region, this part of the region is composed mainly of critical points, which are not cleaned by the agents. Hence, when traveling through this part, the cleaning efficiency of the agents diminishes temporarily, and as the region continues to spread, its global size increases

It is easy to see that after this projection each tile of *slanted-bounding-rectangle(F)* will be "met" by at least a single tile of *F*. In addition, it is impossible that after the projection, two tiles of *slanted-bounding-rectangle(F)* will merge, since for some projecting direction, there are at most one tile of each projection coordinate (namely, when projecting left there is at most one tile for each *Y* coordinate, when projecting downwards there is at most one tile for each *X* coordinate, etc.). Thus, the number of *4 Neighbors* of *F* is at least the number of tiles in *slanted-bounding-rectangle(F)*, denote by *slanted-bounding-rectangle(F)*, namely:

$$\forall F \quad 4 \, Neighbors(F) \geq |slanted\text{-}bounding\text{-}rectangle \ (F)| \tag{3}$$

Note that *slanted-bounding-rectangle* (F) in Fig. 21 contains two pairs of tiles which are 4 Connected.

Definition 8 For each slanted rectangle *R* let us define *canonical-slanted-rectangle (R)* to be the smallest slanted rectangle which bounds *R* such that *canonical-slanted-*

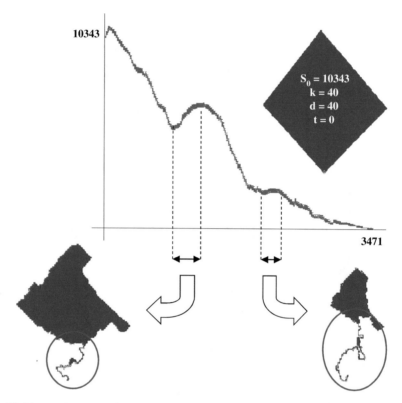

Fig. 19 The graph presents the cleaning process of a (relatively large) "digital sphere". Observe how the size of the region increases at various times. Examining the appearance of the region at those times reveals that the majority of agents are being located at the vicinity of the initial location of the agents (a point which is artificially kept critical, according to the **SWEEP** protocol) — see snapshots at the bottom. After several traversals of the region, this part of the region is composed mainly of critical points, which are not cleaned by the agents. Hence, when traveling through this part, the cleaning efficiency of the agents diminishes temporarily, and as the region continues to spread, its global size increases

$rectangle(R)$ does not contain pairs of tiles which are 4 Connected. In other words, assuming that the world is a chessboard colored in black and white, then $canonical\text{-}slanted\text{-}rectangle(R)$ is the minimal slanted rectangle that contains R, whose tiles have the same color.

It can easily be seen that for some slanted rectangle R, in order to produce:

$$CR = canonical\text{-}slanted\text{-}rectangle(R)$$

at most two sides of the four sides of R must be moved by one tile. In addition, each time a side of R is moved, its length should be increased by at most one. As a result, the number of tiles in CR is at most the number of tiles in R plus 2, namely:

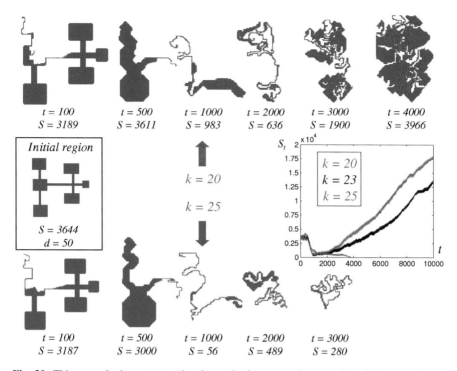

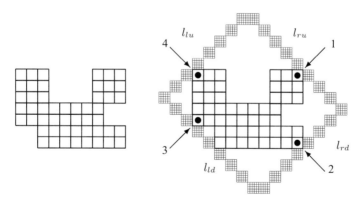

Fig. 20 This example demonstrates the change in the geometric properties of the contaminated region as a result of the cleaning process

Fig. 21 An example of a *slanted-bounding-rectangle*

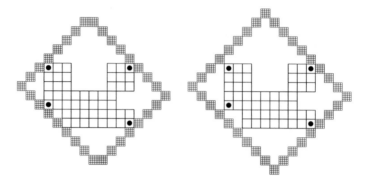

Fig. 22 For a shape F, the right chart demonstrates *slanted-bounding-rectangle(F)* while the left chart demonstrates *canonical-slanted-rectangle(slanted-bounding-rectangle(F))*

$$\forall rectangle\ R \quad |canonical\text{-}slanted\text{-}rectangle(R)| \le |R| + 2 \qquad (4)$$

An example appears in Fig. 22.
Combining Eqs. 3 and 4 we get:

$$\forall F \quad 4\ Neighbors(F) \ge |canonical\text{-}slanted\text{-}bounding\text{-}rectangle(F)| - 2 \qquad (5)$$

Let CR be the smallest canonical slanted rectangle which contains at least S tiles. Let a and b denote the sides of CR and let p denote the number of tiles CR comprises. Then:

$$p = 2(a + b) - 4 \qquad (6)$$

which also means that:

$$a = \frac{p + 4}{2} - b$$

Let $f(a, b)$ denote the area of a canonical slanted rectangle of sides a and b. Since a canonical slanted rectangle contains $(a - 1)$ slanted rows of $(b - 1)$ tiles and $(a - 2)$ slanted rows of $(b - 2)$ tiles we see that:

$$f(a, b) = (a - 1)(b - 1) + (a - 2)(b - 2) \qquad (7)$$

We would now like to find a solution for the following optimization problem:

$$\min \quad p \quad s.t \quad f(a, b) \ge S \quad \wedge \quad p = 2(a + b) - 4$$

After some arithmetics Eq. 7 can be written as:

$$a = \frac{f(a, b) - 5 + 3b}{2b - 3} \qquad (8)$$

Combining this with Eq. 6 we get:

$$p = 2 \cdot \frac{f(a, b) - 5 + 3b}{2b - 3} + 2b - 4$$

Since we require that $f(a, b) \geq S$ we can write the following:

$$p \geq \rho \triangleq 2 \cdot \frac{S - 5 + 3b}{2b - 3} + 2b - 4 \tag{9}$$

Note that the value of b which will minimize ρ (denoted by b_{min}) may not be an integer value, where the meaning of b is the length of the side of CR, which must be an integer number. However, since $\forall b \in \mathbb{N}$ $\rho(b) \geq \rho(b_{min})$ and since $p \geq \rho$, the validity of the bound is preserved (however, the bound may become slightly less tight). In order to minimize ρ we shall calculate:

$$\frac{\partial \rho}{\partial b} = 2 \cdot \frac{3(2b - 3) - 2(S - 5 + 3b)}{(2b - 3)^2} + 2 = 0$$

and after some arithmetics we get that:

$$b = \frac{\sqrt{2S - 1} + 3}{2}$$

By examining the behavior of $\frac{\partial^2 \rho}{\partial b^2}$ we can see that for $b = \frac{\sqrt{2S-1}+3}{2}$ since $S \geq 1$ then $\frac{\partial^2 \rho}{\partial b^2} > 0$, meaning that ρ is indeed minimized at this point.

By assigning the value of b_{min} to Eqs. 8 and 9 we can see that for b_{min}, $a = b$ (meaning that CR is the shape of the perimeter of a digital sphere) and that:

$$p \geq 2(\sqrt{2S - 1} + 1)$$

It is easy to see that a sphere of size S has exactly $2(\sqrt{2S - 1} + 1)$ 4 *Neighbors* and therefore it is the shape that minimizes the number of 4 *Neighbors* for a given area S.

Since the bound was produced for a *canonical* slanted rectangle, and combined with Eq. 5, we get that:

$$\forall F \text{ of size } S \quad 4 \text{ Neighbors}(F) \geq 2\sqrt{2S - 1}$$

8 Discussion and Conclusion

In this chapter the *Dynamic Cooperative Cleaners* problem, where a group of simple and limited robotic agents must scan a spreading "contaminated" sub-grid, was discussed. This problem has several interesting applications, some of which are for example an autonomous patrolling of a pre-defined area using a swarm of collabo-rative drones [8, 12, 17], or an implementation of a distributed anti-virus system for computer networks [15, 24]. Additional applications are distributed search engines [20], as well as the coordination of fire-fighting units (see [21] for details).

Theoretical lower bounds for the cleaning time of a given contaminated region were shown, yielding a way to design input scenarios which are impossible to clean. The cooperative cleaning protocol **SWEEP** was suggested for coping with the dynamic cleaners problem as well, and its performance was compared to this lowers bound.

As in the case of the static variant of the problem, the question of how fast can k drones scan a pre-defined expanding region, is *NP-hard* (for any number of drones, as this can be reduced to finding a Hamilton path in a non-simple grid-graph [23]). Complexity analysis for this can be found in [5, 6], various effects on the completion time as a function of the region's geometric properties are analyzed in [9, 13], and upper bounds are discussed in [16]. Stochastic variant of this problem, involving a probabilistic analysis of the search problem in domains that spread nondeterministi-cally is discussed in [32].

Appendix – Definitions

This Appendix contains terms which were defined in previous chapters, as well as several algorithms and proofs, reiterated for the purpose of self-containment.

The SWEEP Cleaning Protocol

The **SWEEP** protocol is implemented by each agent a_i, located at time t at $\tau_i(t) = (x, y)$. We define below several terms we use while discussing the **SWEEP** protocol. We stress the fact that this is indeed a *myopic* protocol, relying on neighborhood information, a *senile* protocol (the memory needed is constant + one counter whose size is upper bounded by $O(\log k)$, where k is the number of agents) and relies on *implicit local communication* only.

Definition 9 Let ∂F denote the boundary of F. A tile is on the boundary if and only if at least one of its $8Neighbors$ is not in F, meaning:

$$\partial F = \{v \mid v \in F \wedge 8Neighbors(v) \cap (G \setminus F) \neq \emptyset\}$$

Definition 10 Let $\tilde{\tau}_i(t)$ denote the previous location of agent i with respect to $\tau_i(t)$, such that $\tilde{\tau}_i(t) \neq \tau_i(t)$, defined as:

$$\tilde{\tau}_i(t) \triangleq \tau_i(x) \ s.t. \ x = \max\{j \in \mathbb{N} \mid j < t \ \text{and} \ \tau_i(j) \neq \tau_i(t)\}$$

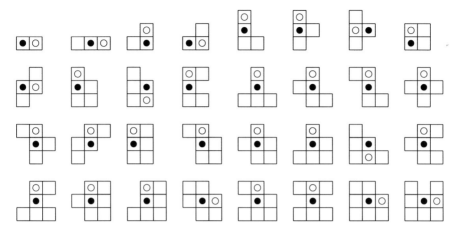

Fig. 23 When $t = 0$ the first movement of an agent located in (x, y) should be decided according to initial contamination status of the neighbors of (x, y), as appears in these charts — the agent's initial location is marked with a *filled circle* while the destination is marked with an *empty one*. All configurations which do not appear in these charts can be obtained by using rotations. This definition is needed in order to initialize the traversal behavior of the agents in the correct direction

Definition 11 The term '*rightmost*' means:

Starting from $\tilde{\tau}_i(t)$ (namely, the previous boundary tile that agent a_i had been in) scan the *four neighbors* of $\tau_i(t)$ in a clockwise order until a boundary tile (excluding $\tilde{\tau}_i(t)$) is found. Sometimes, $\tilde{\tau}_i(t)$ might not be a boundary tile (this may occasionally happen after a contamination spread, for dirty regions of specific geometric features. In those cases, an agent may find itself in an *internal point* — a dirty tile which is not located on the boundary. From this tile the agent must move to the adjacent boundary tile, and resume its traversal along the region's boundary). In this case, if $\tau_i(t)$ is a boundary point, then starting from $\tilde{\tau}_i(t)$ scan the *four neighbors* of $\tau_i(t)$ in a clockwise order until the second boundary tile is found. In case $t = 0$ select the tile as instructed in Fig. 23.

The reference to *contamination spread* used in Definition 11 is given with consideration to the use of this term in the next chapters.

The additional information needed for the protocol and its sub-routines is contained in $\mathcal{M}_i$ and $Neighborhood(x, y)$.

A schematic flowchart of the protocol, describing its major components and procedures is presented in Fig. 24. The complete pseudo-code of the protocol and its sub-routines appears in Figs. 25 and 26. Upon initialization of the system, the *System Initialization* procedure is called (defined in Fig. 25). This procedure sets various initial values of the agents, and call the protocol's main procedure — *SWEEP* (defined in Fig. 26). This procedure in turn, uses various sub-routines and functions, all defined in Fig. 25.

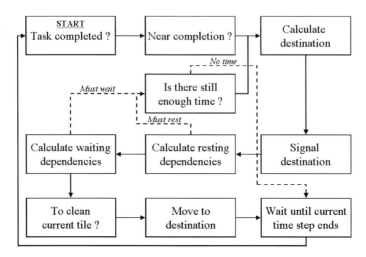

Fig. 24 A schematic flow chart of the **SWEEP** protocol. The *smooth lines* represent the basic flow of the protocol while the *dashed lines* represent cases in which the flow is interrupted. Such interruptions occur when an agent calculates that it must not move until other agents do so (either as a result of *waiting* or *resting* dependencies — see Lines 6 and 19 of **SWEEP** for more details)

Pivot Point

The initial location of the agents (the *pivot point*, denoted as p_0) is artificially set to be critical during the execution, hence it is also guaranteed to be the last point cleaned. Completion of the mission can therefore be concluded from the fact that all (working) agents are back at p_0 with no contaminated neighbors to go to, thereby reporting on completion of their individual missions. Note that this artificial preservation of the criticalness of the pivot point is not necessary for the algorithm to work. Rather, it just makes life easier for the user, as one can now know where to find the agents after the cleaning has been completed. If we do not start with all agents at the pivot and force p_0 to be critical, the location of the agents upon completion will generally not be known in advance.

Signaling

Since by assumption the agent's sensors can detect the status of all tiles which are contained within a digital sphere of radius four placed around the current location of the agent, each agent can artificially calculate the desired destination of all the agents which are located in one of its 4 *Neighbors* tiles (see Fig. 27. Thus, the *signaling* action of each agent can be simulated by the other agents near him, and hence an explicit signaling by the agents is not actually required. However, the signaling action is kept in the description and flowchart of the protocol (in procedure 5) for the sake of simplicity and understandability.

```
 1:  System Initialization
 2:      Arbitrarily choose a pivot point p_0 in ∂F_0, and mark it as critical point
 3:      Place all the agents on p_0
 4:      For (i = 1; i ≤ k; i + +) do
 5:          Call Agent Reset for agent i
 6:          Call SWEEP for agent i
 7:          Wait two time steps
 8:      End for
 9:  End procedure

10:  Agent Reset
11:      resting ← false
12:      dest ← null /* destination */
13:      near completion ← false
14:      saturated perimeter ← false
15:      waiting ← ∅
16:  End procedure

17:  Priority
18:      /* Assuming the agent moved from (x_0, y_0) to (x_1, y_1) */
19:      priority ← 2(x_1 − x_0) + (y_1 − y_0)
20:  End procedure

21:  Check Completion of Mission
22:      If ((x, y) = p_0) and (x, y) has no dirty neighbors then
23:          If (x, y) is dirty then
24:              Clean (x, y)
25:          STOP
26:  End procedure

27:  Check "Near Completion" of Mission
28:      /* Cases where every tile in F_t contains at least a single agent */
29:      near completion ← false
30:      If each of the dirty neighbors of (x, y) contains at least one agent then
31:          near completion ← true
32:      If each of the dirty neighbors of (x, y) has near completion = true then
33:          Clean (x, y) and STOP
34:      /* Cases where every non-critical tile in ∂F_t contains at least 2 agents */
35:      saturated perimeter ← false
36:      If ((x, y) ∈ ∂F_t) and both (x, y) and all of its non-critical neighbors
           in ∂F_t contain at least two agents then
37:          saturated perimeter ← true
38:      If ((x, y) ∈ ∂F_t) and both (x, y) and all of its neighbors in ∂F_t has
           saturated perimeter = true then
39:          Ignore resting commands for this time step
40:  End procedure
```

Fig. 25 The first part of the **SWEEP** cleaning protocol

Connectivity Preservation

The connectivity of the region yet to be cleaned, F, is preserved by allowing the agents to clean only *non-critical* points. This guarantees both successful termination and agreement of completion (since having no dirty neighbors implies that $F = \emptyset$). Also, should several agents malfunction and halt, as long as there are still functioning agents, the mission will still be carried out, albeit slower.

1: **SWEEP Protocol** /* Controls agent i after **Agent Reset** */
2:      **Check Completion of Mission**
3:      **Check "Near Completion" of Mission**
4:      *dest ← rightmost neighbor* of (x,y) /* Calculate destination */
5:      *destination signal bits ← dest* /* Signaling the desired destination */
6:      /* Calculate resting dependencies (solves agents' clustering problem) */
7:      Let all agents in (x,y) except agent i be divided to the following groups :
8:          A_1 : Agents signaling towards any direction different than *dest*
9:          A_2 : Agents signaling towards *dest* which entered (x,y) before agent i
10:         A_3 : Agents signaling towards *dest* which entered (x,y) after agent i
11:         A_4 : Agents signaling towards *dest* which entered (x,y) with agent i
12:     Let group A_4 be divided into the following two groups :
13:         A_{4a} : Agents with lower *priority* than this of agent i
14:         A_{4b} : Agents with higher *priority* than this of agent i
15:     *resting ← false*
16:     **If** $(A_2 \neq \emptyset)$ or $(A_{4b} \neq \emptyset)$ then
17:         *resting ← true*
18:         **If** (current time-step $\mathcal{T}$ did not end yet) then jump to 4 **Else** jump to 35
19:     *waiting ← $\emptyset$* /* Waiting dependencies (agents synchronization) */
20:     Let *active agent* denote a *non-resting* agent which didn't move in $\mathcal{T}$ yet
21:     **If** (x-1,y) $\in F_t$ contains an active agent then *waiting ← waiting* $\cup$ {*left*}
22:     **If** (x,y-1) $\in F_t$ contains an active agent then *waiting ← waiting* $\cup$ {*down*}
23:     **If** (x-1,y-1) $\in F_t$ contains an active agent then *waiting ← waiting* $\cup$ {*l-d*}
24:     **If** (x+1,y-1) $\in F_t$ contains an active agent then *waiting ← waiting* $\cup$ {*r-d*}
25:     **If** *dest = right* and (x+1,y) contains an active agent j, and *dest$_j$* $\neq$ *left*, and
            there are no other agents delayed by agent i (i.e. (x-1,y) does not contain
            active agent l with *dest$_l$* =*right* and no active agents in (x,y+1),(x+1,y+1),
            (x-1,y+1), and (x+1,y) does not contain active agent n with *dest$_n$* = *left*),
            then $($*waiting ← waiting* $\cup$ {*right*}$)$ and $\left(\textit{waiting}_j \leftarrow \textit{waiting}_j \setminus \{\textit{left}\}\right)$
26:     **If** *dest = up* and (x,y+1) contains an active agent j, and *dest$_j$* $\neq$ *down*, and
            there are no other agents delayed by agent i (i.e. (x,y-1) does not contain
            active agent l with *dest$_l$* =*up* and no active agents in (x+1,y),(x+1,y+1),
            (x-1,y+1), and (x,y+1) does not contain active agent n with *dest$_n$* = *down*),
            then $($*waiting ← waiting* $\cup$ {*up*}$)$ and $\left(\textit{waiting}_j \leftarrow \textit{waiting}_j \setminus \{\textit{down}\}\right)$
27:     **If** (*waiting* $\neq \emptyset$) then
28:         **If** ($\mathcal{T}$ has not ended yet) then jump to 4 **Else** jump to 35
29:     /* Decide whether or not (x,y) should be cleaned */
30:     **If** $\neg$ ((x,y) $\in \partial F_t$) or ((x,y) $\equiv p_0$) or (x,y) has 2 dirty tiles in its $4Neighbors$
            which are not connected via a path of dirty tiles from its $8Neighbors$ then
31:         (x,y) is an *internal point* or a *critical point* and should not be cleaned
32:     **Else**
33:         Clean (x,y) if and only if it does not still contain other agents
34:     Move to *dest*
35:     Wait until $\mathcal{T}$ ends.
36:     Return to 2

Fig. 26 The **SWEEP** cleaning protocol. The term *rightmost neighbor* is defined in Appendix section "The SWEEP Cleaning Protocol". *l-d* and *r-d* are *left-down* and *right-down*, respectively

Fig. 27 Digital sphere of diameter 7, placed around the agent

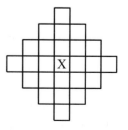

Note that a tile which was originally clean, or which was cleaned by the agents, is guaranteed to remain clean. As a tile can be cleaned by the agents only when it is in ∂F, it is easy to see that the simple connectivity of F is maintained throughout the cleaning process (as the creation of "holes" — clean tiles which are surrounded by tiles of F — is impossible).

Note that when several agents are located in the same tile, only the last one who exits cleans it (see Line 33 in **SWEEP**), in order to prevent agents from being 'thrown' out of the dirty region (meaning, cleaning a tile in which another agent is located and thus preventing this agent from being able to continue the execution of the cleaning protocol). An alternative method for ensuring the agents are always capable of executing the cleaning protocol would have been the implementation of a "dirtiness seeking mechanism" (i.e. by applying methods such as suggested in [18]). Such a mechanism would have allowed an agent to clean its current location, even if other agents had also been present there. That solution, however, would have been far less elegant and would have added additional difficulties to the analysis process.

Agents Synchronization

Note that agents operating in the described environment must have some mean of synchronization, which is mandatory in order to prevent agents from operating in the same time — risking to cut the dirty region into several connected components, as shown in Fig. 28.

To ensure such scenarios will not occur, an order between the operating agents must be implemented. Note — throughout the next paragraphs agents which are signaling a *resting* status (see Appendix section "Clustering Problem" for more details) are being disregarded while calculating the dependencies of the agents' movements. The creation of the following order is implemented in Procedure 19 of **SWEEP**:

Fig. 28 When the agents do not possess a synchronization mechanism, they may, among others, damage the region's connectivity. In this example, two agents clean their current locations, and move according to the **SWEEP** protocol. Since they are not synchronized, the tiles which they are located in are not treated as critical at the time of cleaning. However, the region's connectivity is not preserved. Should one of the agents had waited for its neighbor to complete executing the protocol's steps before resuming its actions, while deciding whether to clean its current location, it would have treated this tile as critical, and therefore avoid cleaning it. In this case, the connectivity of the region would have been maintained

Definition 12 For agent i, let $P_i \subseteq \{up, down, left, right, right\text{-}down, left\text{-}down\}$ be a set of directions of tiles, in which there are currently agents, which agent i is delayed by (meaning, agent i will not start moving until the agents in these tiles move). Unless stated otherwise, $P_i = \emptyset$.

For agent i which is located in tile (x, y), if $(x - 1, y)$ is a tile in F, which contains an agent, then $P_i \leftarrow P_i \cup \{left\}$. If $(x, y - 1)$ is a tile in F, which contains an agent, then $P_i \leftarrow P_i \cup \{down\}$, and similarly for $(x + 1, y - 1)$ and $right\text{-}down$ and for $(x - 1, y - 1)$ and $left\text{-}down$.

Definition 13 Let $dest_i \in \{up, down, left, right\}$ be the destination agent i might be interested in moving to, after leaving its current location.

Let each agent be equipped with a two bit flag (that may be implemented for example by two small light-bulbs). This flag states the desired destination of the agent. Alternatively, each tile can be treated as a physical tile, in which the agent can move. Thus, the agent can move towards the top side of the tile, which will be equivalent for using the flag in order to signal that it intends to move *up*.

Let each time step be divided into two phases. In phase 1, every agent "signals" the destination it intends to move towards, either by moving to the appropriate side of the tile, or by using the destination flag.

As we defined an artificial rule which states the superiority of *left* and *down* over *right* and *up* (and internally, of *down* over *left*), there are several specific scenarios in which this asymmetry should be reversed in order to ensure a proper operation of the agents. Following is such a a "dependencies switching" rule: For agent i, which is located in (x, y), if $dest_i = right$ and tile $(x + 1, y)$ contains an agent j, and $dest_j \neq left$, and there are no other agents which are delayed by agent i (i.e. tile $(x - 1, y)$ does not contain an agent l where $dest_l = right$ and tiles $(x, y + 1), (x + 1, y + 1), (x - 1, y + 1)$ do not contain any agent and tile $(x + 1, y)$ does not contain an agent n where $dest_n = left$), then $P_i \leftarrow P_i \cup \{right\}$ and $P_j \leftarrow P_j \setminus \{left\}$. Also, if $dest_i = up$ and tile $(x, y + 1)$ contains an agent k, and $dest_k \neq down$, and there are no other agents which are delayed by agent i (i.e. tile $(x, y - 1)$ does not contain an agent m where $dest_m = up$ and tiles $(x + 1, y), (x + 1, y + 1), (x - 1, y + 1)$ do not contain any agent and tile $(x, y + 1)$ does not contain an agent q where $dest_q = down$), then $P_i \leftarrow P_i \cup \{up\}$ and $P_k \leftarrow P_k \setminus \{down\}$.

At phase 2 of each time step, the agents start to operate in turns, according to the order implied by P_i. This guarantees that the connectivity of the region is kept, since the simultaneous movement of two neighboring agents is prevented.

Notice that deadlocks are impossible — since the basic rule is that every agent is delayed by the agents in its *left* and *down* neighbor tiles. Therefore, at any given time, and for every possible group of agents, there exist an agent with the minimal x and y coordinates (which by definition, is not delayed by any other agent of this group). After this agent moves, all the agents which are delayed by it, can now move, and so on. As to the "dependencies switching rule" — let agent i located in tile (x, y) have the minimal x and y values among the agents who had not moved yet, and let

$dest_i = up$, and let tile $(x, y+1)$ contain an agent j such that $dest_j \neq down$. Then although agent i is located below agent j, it will be delayed by it (i.e. $(up \in P_i)$ and $\neq (down \in P_j)$) as long as agent i is not delaying any other agent (as this is the requirement of the "dependencies switching rule"). In this case, we should show that there can not be a cycle of dependencies, which starts at agent j, ends at agent i, and is closed by the dependency of agent i on agent j. Such a circle can not exist since for it to end in agent i, it means that agent i is delaying another agent k. However, this is impossible since agent i is known not to delay any agent (specifically agent k). Hence, circular dependencies are prevented and no deadlock are possible.

Note that while phase 2 is in process, F_t may change due to cleaning by the agents. As a result, the desired destinations of the agents as well as their dependencies forest, must also be dynamic. This is achieved through the repeated recalculation of these values, for every $waiting$ agent. For example, assume agent i to be located in (x, y), and $dest_i = down$ without loss of generality, and let agent j located in tile $(x, y-1)$ moves out of this tile and cleans it. Then, $dest_i$ naturally change (as the tile (x, y) does no longer belong to F_t, and thus is not a legitimate destination for agent i), and thus, P_i may also change. In this case, agent i should change its "destination signal" and act according to its new $dest_i$ and P_i. This is implemented in **SWEEP** by calculating the waiting agents' destinations and dependencies lists repeatedly, until either all agents have moved or until the time step has ended (meaning that some agents had to change their status to $resting$, and pause until the next time step — see Appendix section "Clustering Problem" for more details). Note that every $waiting$ agent is guaranteed either to complete its movement in the current time step, or to be forced to wait for the next time step, by switching its status to $resting$.

Notice that the "dependencies switching rule" is not required in order to ensure a proper completion of the mission, but rather to improve the agents' performance, by preventing a bug demonstrated in Fig. 29.

Clustering Problem

Since we are interested in preventing the agents from moving together and functioning as a single agent (due to a resulting decrease in the system's performance), we would like to impose a "resting policy" which will ensure that the agents do not group into clusters which move together. This is done by artificially delaying agents in their current locations, should several agents placed in the same tile wish to move to the same destination. However, we would like the resting time of the agents to be minimal, for performance and analysis reasons.

The following resting policy is implemented by Sect. 6 of **SWEEP**. Using its sensors, an agent intending to move into tile v is aware of other agents which intend to move into v as well. Thus, when an agent enters a tile which already contains other agents, it knows whether these agents entered this tile at the same time step it did, or whether they had occupied this tile before the current time step had started.

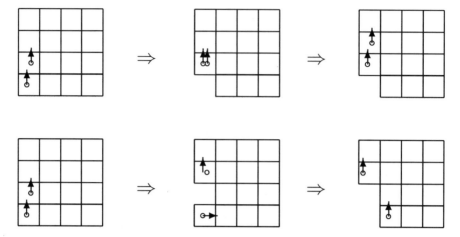

Fig. 29 The upper charts demonstrate a performance bug which may be caused due to local dependencies. The agents are advancing according to the **SWEEP** protocol, but their cleaning performance is decreased. The lower charts demonstrates the cleaning operation after adding the dependencies switching mechanism

Note that in phase 1 of each time step all the agents are signaling their desired destinations. Thus, an agent can identify which ones of the agents which are located in its current tile intend to move to the same destination as its. Only those agents may cause this agent to delay its actions and rest.

From the agents which intend to move to the same direction as agent i, the agent can distinguish between three groups of agents — the agents which entered this tile before the time step agent i did (group A_2), those which entered the tile in the same time step as i (group A_4), and those who entered it after agent i did (group A_3). If $A_2 \neq \emptyset$, agent i waits, change its status to *resting agent*, signals this status to the other agents in its vicinity, and does not move. As this rule is kept by every other agent, agent i is guaranteed that the agents of group A_3 in their turn will wait for agent i moves before being free to do so as well.

As to the agents of A_4, the problem is solved by induction over the time steps, namely — that at any given time, two agents can not leave the same tile towards the same direction at the same time step. The base of this induction holds for small values of t as the agents are periodically released by the initialization procedure of **SWEEP**, while no two agents are released at the same time step. Later, as we are assured that all the agents of group A_4 arrived to v from different tiles (which are also different than the tile agent i had entered v from), then since all the agents in group A_4 know the previous locations of each other, a consensus over a local ordering of A_4 is established (note that no explicit communication is needed to form this ordering). An example for such an order is the **Priority** function of the **SWEEP** protocol. As a result, the agents are able to exit the tile they are currently located in, in an orderly fashion, according to a well defined order. Thus, the following invariant holds: "*at* *any given time t, for any two tiles v, u, there can only be a single agent which moves from v to u at time step t"*. Thus, the clustering problem is solved.

Notice that there is a single exception to this mechanism, in which the resting commands are overruled. This happens when all non-critical perimeter tiles contain at least two agents. In this scenario, following the resting commands would have created a situation in which no tile can be cleaned, as for every agent which leaves a certain tile, there are still "resting" agents located in the same tile. Therefore, once this scenario is detected by the *check near completion of mission* procedure of **SWEEP**, the agents ignore their resting commands momentarily, in order to be able to clean the non-critical perimeter tiles. The order and internal prioritization of the agents are maintained, for calculating the new *resting* commands in the following time step.

Mission Termination

The termination of the protocol is done in one of two cases — either an agent finds itself in the pivot point, while all of its neighbors are clean (which means that this is the last dirty tile and therefore should be cleaned), or when each dirty tile contains at least one agent (which is a generalization of the previous scenario). The second case is implemented by allowing the agents to signal to their four neighbors whether all of their dirty neighbors contain at least a single agent.

References

1. M. Ahmadi, P. Stone, A multi-robot system for continuous area sweeping tasks, in *IEEE International Conference on Robotics and Automation, 2006 (ICRA 2006)* (2006)
2. S. Alpern, S. Gal, *The Theory of Search Games and Rendezvous* (Kluwer Academic Publishers, Boston, 2003)
3. Y. Altshuler, *Multi Agents Robotics in Dynamic Environments*. Ph.D. thesis, Israeli Institute of Technology, May 2010
4. Y. Altshuler, A. Bruckstein, On short cuts-or-fencing in rectangular strips (2010), arXiv:1011.5920
5. Y. Altshuler, A.M. Bruckstein, The complexity of grid coverage by swarm robotics, in *ANTS 2010* (LNCS, 2010), pp. 536–543
6. Y. Altshuler, A.M. Bruckstein, Static and expanding grid coverage with ant robots: Complexity results. Theoret. Comput. Sci. **412**(35), 4661–4674 (2011)
7. Y. Altshuler, A.M. Bruckstein, I.A. Wagner, Swarm robotics for a dynamic cleaning problem, in *IEEE Swarm Intelligence Symposium* (2005), pp. 209–216
8. Y. Altshuler, V. Yanovski, I.A. Wagner, A.M. Bruckstein, The cooperative hunters - efficient cooperative search for smart targets using uav swarms, in *Second International Conference on Informatics in Control, Automation and Robotics (ICINCO), the First International Workshop on Multi-Agent Robotic Systems (MARS)* (2005), pp. 165–170
9. Y. Altshuler, I.A. Wagner, A.M. Bruckstein, Shape factor's effect on a dynamic cleaners swarm, in *Third International Conference on Informatics in Control, Automation*

and Robotics (ICINCO), the Second International Workshop on Multi-Agent Robotic Systems (MARS) (2006), pp. 13–21

10. Y. Altshuler, V. Yanovsky, I. Wagner, A. Bruckstein, Swarm intelligence searchers, cleaners and hunters, *Swarm Intelligent Systems* (2006), pp. 93–132

11. Y. Altshuler, V. Yanovski, D. Vainsencher, I.A. Wagner, A.M. Bruckstein, On minimal perimeter polyminoes, in *The 13th International Conference on Discrete Geometry for Computer Imagery (DGCI2006)* (2006), pp. 17–28

12. Y. Altshuler, V. Yanovsky, A.M. Bruckstein, I.A. Wagner, Efficient cooperative search of smart targets using uav swarms. ROBOTICA **26**, 551–557 (2008)

13. Y. Altshuler, I.A. Wagner, A.M. Bruckstein, On swarm optimality in dynamic and symmetric environments. Economics **7**, 11 (2008)

14. Y. Altshuler, I.A. Wagner, A.M. Bruckstein, Collaborative exploration in grid domains, in *Sixth International Conference on Informatics in Control, Automation and Robotics (ICINCO)* (2009)

15. Y. Altshuler, S. Dolev, Y. Elovici, N. Aharony, *Ttled random walks for collaborative monitoring, in NetSciCom, (Second International Workshop on Network Science for Communication Networks), San Diego*, vol. 3 (CA, USA, 2010)

16. Y. Altshuler, I.A. Wagner, V. Yanovski, A.M. Bruckstein, Multi-agent cooperative cleaning of expanding domains. Int. J. Robot. Res. **30**, 1037–1071 (2010)

17. Y. Altshuler, A. Pentland, S. Bekhor, Y. Shiftan, A. Bruckstein, Optimal dynamic coverage infrastructure for large-scale fleets of reconnaissance uavs (2016), arXiv:1611.05735

18. R. Baeza-Yates, R. Schott, Parallel searching in the plane. Comput. Geom. **5**, 143–154 (1995)

19. R. Bejar, B. Krishnamachari, C. Gomes, B. Selman, Distributed constraint satisfaction in a wireless sensor tracking system, in *Proceedings of the IJCAI-01 Workshop on Distributed Constraint Reasoning* (2001)

20. M. Bender, S. Michel, P. Triantafillou, G. Weikum, C. Zimmer, Minerva: collaborative p2p search, in *Proceedings of the 31st international conference on Very large data bases* (2005), pp. 1263–1266

21. D.W. Casbeer, D.B. Kingston, R.W. Beard, T.W. McLain, Cooperative forest fire surveillance using a team of small unmanned air vehicles. Int. J. Syst. Sci. **37**(6), 351–360 (2006)

22. Y. Gabriely, E. Rimon, Competitive on-line coverage of grid environments by a mobile robot. Comput. Geom. **24**, 197–224 (2003)

23. A. Itai, C.H. Papadimitriou, J.L. Szwarefiter, Hamilton paths in grid graphs. SIAM J. Comput. **11**, 676–686 (1982)

24. R. Janakiraman, M. Waldvogel, Q. Zhang, Indra: a peer-to-peer approach to network intrusion detection and prevention, in *Proceedings of IEEE WETICE* (2003)

25. W. Kerr, D. Spears, Robotic simulation of gases for a surveillance task, in *Intelligent Robots and Systems (IROS 2005)* (2005), pp. 2905–2910

26. S. Koenig, Y. Liu, Terrain coverage with ant robots: a simulation study, in *AGENTS'01* (2001)

27. B.O. Koopman, The theory of search ii, target detection. Oper. Res. **4**(5), 503–531 (1956)

28. D. Latimer, S. Srinivasa, V. Lee-Shue, S. Sonne, H. Choset, A. Hurst, Toward sensor based coverage with robot teams, in *Proceedings of the 2002 IEEE International Conference on Robotics and Automation* (2002)

29. T.W. Min, H.K. Yin, A decentralized approach for cooperative sweeping by multiple mobile robots, in *IEEE/RSJ International Conference on Intelligent Robots and Systems* (1998), pp. 380–385

30. K. Passino, M. Polycarpou, D. Jacques, M. Pachter, Y. Liu, Y. Yang, M. Flint, M. Baum, *Cooperative Control for Autonomous Air Vehicles, chapter Cooperative Control and Optimization* (Kluwer Academic, Boston, 2002)

31. M. Polycarpou, Y. Yang, K. Passino, A cooperative search framework for distributed agents, in *IEEE International Symposium on Intelligent, Control* (2001), 1–6

32. E. Regev, Y. Altshuler, A.M. Bruckstein, The cooperative cleaners problem in stochastic dynamic environments (2012), arXiv:1201.6322

33. I. Rekleitis, V. Lee-Shuey, A. Peng Newz, H. Choset, Limited communication, multi-robot team based coverage, in *IEEE International Conference on Robotics and Automation* (2004)
34. L.D. Stone, *Theory of Optimal Search* (Academic Press, New York, 1975)
35. J. Svennebring, S. Koenig, Building terrain-covering ant robots: a feasibility study. Auton. Robots **16**(3), 313–332 (2004)
36. D. Vainsencher, A.M. Bruckstein, On isoperimetrically optimal polyforms. Theor. Comput. Sci. **406**(1–2), 146–159 (2008)
37. I.A. Wagner, Y. Altshuler, V. Yanovski, A.M. Bruckstein, Cooperative cleaners: a study in ant robotics. Int. J. Robot. Res. (IJRR) **27**(1), 127–151 (2008)

Swarm Search of Expanding Regions in Grids: Upper Bounds

1 Introduction

One of the most interesting challenges for a swarm of robotic drones is the design and analysis of a multi-robotics system for searching areas (whose dimensions and shape are either known or unknown) [18, 27, 28, 32, 35, 37] or see [3] for a survey of search and evasion strategies. While in most works the targets of the search mission were assumed to be idle, recent works considered dynamic targets, meaning — targets which after being detected by the searching robots, respond by performing various evasive maneuvers intended to prevent their interception.

In this chapter we discuss the problem concerning a given (yet potentially unknown) region that needs to be searched (or *"cleaned"*) by a swarm of autonomous robotic agents, namely – to have all of its 'tiles', or 'pixels' visited at least once by at least a single member of the swarm, and that this "cleaning" would be guaranteed to be completed within a finite (and as short as possible) time. We further assume that the region that needs to be searched *"expands"* over time, namely – that parts of it that were already "cleaned" (or searched) can become "contaminated" (or "un-searched") again. Under this assumption, the completion of the cleaning mission is achieved only when all of the region's pixels are in "clean" status (e.g. were visited by the drones, and did not become contaminated again yet). This dynamic variant of the problem assumes a deterministic evolution of the environment, simulating a spreading *contamination*, or *fire* [7].

As already stated in previous chapters, since we know no easy way to decide whether k agents can successfully clean an instance of the *Dynamic Cooperative Cleaners* problem, producing bounds for the proposed cleaning protocol is important for estimating its efficiency. In this chapter we discuss the efficiency of the proposed **SWEEP** cleaning protocol (discussed in the previous chapters, as well as in [39]), by analyzing its time-complexity and producing upper bounds on the cleaning time of a group of k agents.

This chapter is based on work previously published in parts in [4, 10, 13].

© Springer International Publishing AG 2018
Y. Altshuler et al., *Swarms and Network Intelligence in Search*,
Studies in Computational Intelligence 729, DOI 10.1007/978-3-319-63604-7_4

We assume that the swarm is collaborative, namely – that the various members of it are equipped with the same (and properly synchronized) software, and that each robot can acquire only the information which is available in its immediate vicinity, with the only way of inter-robot communication is by leaving traces on the common ground and sensing the traces left by other robots. Similar works can be found at [2, 22, 29, 30, 33].

This chapter is organized as follows: The problem's definitions appear in Sect. 2. The collaborative "swarm cleaning protocol" is presented in Sect. 3. A recursive upper bound over the cleaning time of a group of drones appears in Sect. 4, while various direct approximations for this bound are discussed in Sects. 5 and 6. An upper bound for the minimal number of drones required in order to guarantee a successful cleaning of a given contaminated region in presented in Sect. 7. Concluding remarks and discussion appears in Sect. 8, whereas several terms, mechanisms and proofs that were discussed in previous chapters, but appear here for the purpose of self-containment (see Appendix).

2   The Dynamic "Cooperative Cleaners" Problem

The definition of the dynamic cooperative cleaners problem is very similar to that of the static variant, albeit with several required extensions. For the sake of readability, following is the complete self-contained definition of the problem.

We shall assume that the time is discrete.

Definition 1 Let the undirected graph $G(V, E)$ denote a two dimensional integer grid $\mathbf{Z}^2$, whose vertices (or "*tiles*") have a binary property called '*contamination*'. Let $cont_t(v)$ denote the contamination state of the tile v at time t, taking either the value "*on*" (for "dirty" or "contaminated") or "*off*" (for "clean").

For two vertices $v, u \in V$, the edge (v, u) may belong to E at time t only if both of the following hold:

- v and u are *4Neighbors* in G.
- $cont_t(v) = cont_t(u) = on$.

This however is a necessary but not a sufficient condition.

The edges of E represent the connectivity of the contaminated region. At $t = 0$ all the contaminated tiles are connected, namely:

$$(v, u) \in E_0 \iff (v, u \text{ are } 4Neighbors \text{ in } G) \wedge (cont_0(v) = cont_0(u) = on)$$

Edges may be added to E only as a result of a contamination spread and can be removed only while contaminated tiles are cleaned by the agents (see below).

Definition 2 Let $F_t(V_{F_t}, E_t)$ be the contaminated sub-graph of G at time t, i.e.:

$$V_{F_t} = \{v \in G \mid cont_t(v) = on\}$$

We assume that F_0 is a single simply-connected component (the actions of the agents will be so designed that this property will be preserved).

Let a group of k agents that can move on the grid G (moving from a tile to its neighbor in one time step) be placed at time t_0 on F_0, at point $p_0 \in V_{F_t}$.

Each agent is equipped with a sensor capable of telling the *contamination* status of all tiles in the digital sphere of diameter 7, which surrounds the agent. An agent is also aware of other agents which are located in these tiles, and all the agents agree on a common direction. Each tile may contain any number of agents simultaneously. This information will later be required by the agents' cleaning protocol. Each agent is equipped with a memory of size $O(\log k)$ bits (used later as an agents counter).

When an agent moves to a tile v, it has the possibility of cleaning this tile (i.e. causing $cont(v)$ to become *off*. Once an agent cleaned a tile v, all the edges touching v are removed, namely:

$$E_{t+1} = E_t \setminus \{(v, u) \mid (v, u) \in E_t \wedge cont(v) = off\}$$

The agents do not have any prior knowledge of the shape or size of the sub-graph F_0 except that it is a single and simply connected component.

The contaminated region F_t is assumed to be surrounded at its boundary by a "rubber-like" elastic barrier, dynamically reshaping itself to fit the evolution of the contaminated region over time (the barrier is derived from the edges of E_t, as demonstrated in Fig. 1). This process can be thought of as a water filed rubber balloon — while the shape of the water contained in it may change, as well as its volume the balloon keeps bounding it tightly. This barrier is intended to guarantee

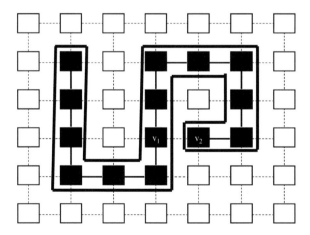

Fig. 1 An illustration of the barrier surrounding the contaminated region, derived from the edges of E_t. Notice that when an edge connecting two contaminated tiles are not in E_t, those tiles are "on the other side" of the barrier. For example, observe the vertices v_1 and v_2, which are 4*Neighbors* in F_t (namely, the Manhattan distance between them in G equals one), however — as $\neg((v_1, v_2) \in E_t)$ the geodesic distance between them is significantly greater

Fig. 2 An illustration of the evolution of the elastic barrier as a result of the movement and cleaning of an agent according to the **SWEEP** protocol, (as described in Figs. 17 and 18)

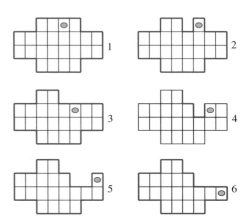

the preservation of the simple connectivity of F_t, in order to allow the agents to perform their cleaning activities. The need for such a barrier is discussed in Sect. 8.

When an agent cleans a contaminated tile, the barrier is withdrawn, in order to free the void previously occupied by the cleaned tile. This is demonstrated in Fig. 2.

Definition 3 Let d denote the number of time steps between contamination spreads. That is, if $t = nd$ for some positive integer n, then:

$$\forall v \in F_t \ \forall u \in 4Neighbors(v) \ , \ cont_{t+1}(u) = on$$

While the contamination spreads, the elastic barrier stretches accordingly. First, all the clean tiles located to the right of a contaminated tile are becoming contaminated. Then, the clean tiles located below contaminated tiles, followed by clean tiles located to the left of such tiles. Finally, clean tiles located above a contaminated tile are affected. The barrier itself it implicitly derived from the edges connecting the tiles. Namely, it is the boundary which surrounds the connected contaminated tiles. This process is illustrated in Figs. 3 and 4, and can be formalized as follows:

Definition 4 Elastic barrier's expansion as a result of a contamination spread:

1. Let $\widehat{V_t} = \emptyset$ be a set of vertices and let $\widehat{E_t} = \emptyset$ be a set of edges
2. For vertex u located at (x_u, y_u) let:

    - $Neighbor_{right}(u)$ denote the vertex located at $(x_u + 1, y_u)$
    - $Neighbor_{left}(u)$ denote the vertex located at $(x_u - 1, y_u)$
    - $Neighbor_{up}(u)$ denote the vertex located at $(x_u, y_u + 1)$
    - $Neighbor_{down}(u)$ denote the vertex located at $(x_u + 1, y_u - 1)$

3. Let $d_G(v, u)$ denote the number of edges between v and u in the graph G
4. Add new vertices to the right —

    (a) $V_{temp} = \left\{ v \mid \neg(v \in V_{F_t} \cup \widehat{V_t}) \wedge \left(\exists u \in V_{F_t} , \ v = Neighbor_{right}(u)\right)\right\}$
    (b) $\widehat{E_t} \leftarrow \widehat{E_t} \cup \left\{(v, u) \mid (v \in V_{temp}) \wedge (v = Neighbor_{right}(u))\right\}$

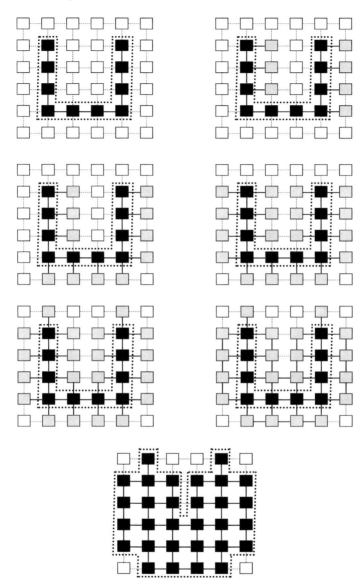

Fig. 3 An illustration of the barrier expansion process as a result of a contamination spread. The example starts from the top left chart, going right

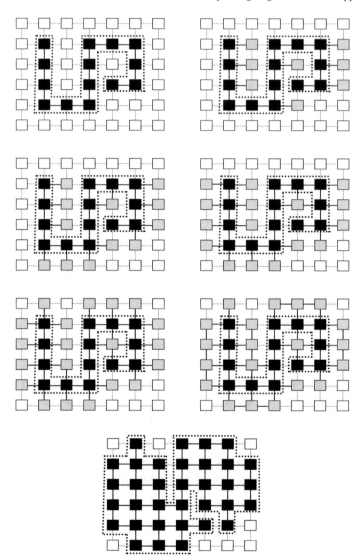

Fig. 4 An illustration of the barrier expansion process as a result of a contamination spread. The example starts from the top left chart, going right

(c) $\widehat{E}_t \leftarrow \widehat{E}_t \cup \left\{ (v, u) \; \middle| \; \begin{array}{l} (v \in \widehat{V}_t) \wedge (v = Neighbor_{right}(u)) \wedge \\ (u \in V_{F_t}) \wedge (d_{G(V_{F_t} \cup \widehat{V}_t, E_t \cup \widehat{E}t)}(v, u) \leq 3) \end{array} \right\}$

(d) $\widehat{V}_t \leftarrow \widehat{V}_t \cup V_{temp}$

5. Add new vertices to the bottom —

    (a) $V_{temp} = \left\{ v \mid \neg(v \in V_{F_t} \cup \widehat{V}_t) \wedge (\exists u \in V_{F_t}, \; v = Neighbor_{down}(u)) \right\}$

    (b) $\widehat{E}_t \leftarrow \widehat{E}_t \cup \left\{ (v, u) \mid (v \in V_{temp}) \wedge (v = Neighbor_{down}(u)) \right\}$

    (c) $\widehat{E}_t \leftarrow \widehat{E}_t \cup \left\{ (v, u) \; \middle| \; \begin{array}{l} (v \in \widehat{V}_t) \wedge (v = Neighbor_{down}(u)) \wedge \\ (u \in V_{F_t}) \wedge (d_{G(V_{F_t} \cup \widehat{V}_t, E_t \cup \widehat{E}t)}(v, u) \leq 3) \end{array} \right\}$

    (d) $\widehat{V}_t \leftarrow \widehat{V}_t \cup V_{temp}$

6. Add new vertices to the left —

    (a) $V_{temp} = \left\{ v \mid \neg(v \in V_{F_t} \cup \widehat{V}_t) \wedge (\exists u \in V_{F_t}, \; v = Neighbor_{left}(u)) \right\}$

    (b) $\widehat{E}_t \leftarrow \widehat{E}_t \cup \left\{ (v, u) \mid (v \in V_{temp}) \wedge (v = Neighbor_{leftt}(u)) \right\}$

    (c) $\widehat{E}_t \leftarrow \widehat{E}_t \cup \left\{ (v, u) \; \middle| \; \begin{array}{l} (v \in \widehat{V}_t) \wedge (v = Neighbor_{left}(u)) \wedge \\ (u \in V_{F_t}) \wedge (d_{G(V_{F_t} \cup \widehat{V}_t, E_t \cup \widehat{E}t)}(v, u) \leq 3) \end{array} \right\}$

    (d) $\widehat{V}_t \leftarrow \widehat{V}_t \cup V_{temp}$

7. Add new vertices to the top —

    (a) $V_{temp} = \left\{ v \mid \neg(v \in V_{F_t} \cup \widehat{V}_t) \wedge (\exists u \in V_{F_t}, \; v = Neighbor_{up}(u)) \right\}$

    (b) $\widehat{E}_t \leftarrow \widehat{E}_t \cup \left\{ (v, u) \mid (v \in V_{temp}) \wedge (v = Neighbor_{up}(u)) \right\}$

    (c) $\widehat{E}_t \leftarrow \widehat{E}_t \cup \left\{ (v, u) \; \middle| \; \begin{array}{l} (v \in \widehat{V}_t) \wedge (v = Neighbor_{up}(u)) \wedge \\ (u \in V_{F_t}) \wedge (d_{G(V_{F_t} \cup \widehat{V}_t, E_t \cup \widehat{E}t)}(v, u) \leq 3) \end{array} \right\}$

    (d) $\widehat{V}_t \leftarrow \widehat{V}_t \cup V_{temp}$

9. $V_{F_t} \leftarrow V_{F_t} \cup \widehat{V}_t$

10. $E_t \leftarrow E_t \cup \widehat{E}_t$

11. $E_t \leftarrow E_t \cup \left\{ (v, u) \mid (v, u \in V_{F_t}) \wedge (v = 4Neighbors(u)) \wedge (d_{F_t}(v, u) \leq 3) \right\}.$

For the agents who travel along the tiles of F, the barrier signals the boundary of the contaminated region. When an agent detects a contaminated tile which is "on the other side" of the barrier it treats this tile as though it was a clean tile (this aspect is important to remember when reviewing the agents' behavior while following the cleaning protocol). For example, examining the region which appears in Fig. 5 and assuming an agent located in the "X" marked tile, then while looking "upwards", this agent acts as though the tile above it is clean (as the contaminated tiles are located behind the barrier and are thus masked as clean ones).

As in the case of the static variant, it is the agents' goal to clean G by eliminating the contamination entirely, meaning that the agents must ensure that:

$$\exists t_{success} \; s.t \; F_{t_{success}} = \emptyset$$

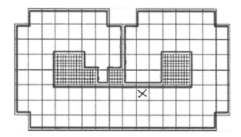

Fig. 5 Observe how the stretching of the barrier is generating "double-fronts" after a contamination spread. Note that had the contaminated parts been allowed to merge, the simple connectivity of the region would have not been kept. In addition, observe how an agent located in the tile marked by an "X" is still kept on the boundary of the contaminated region, due to this mechanism

In addition, it is desired that time $t_{success}$ will be minimal.

In addition, and as stated previously, we still impose the restriction of no central control and a fully 'de-centralization' system.

3   The Cleaning Protocol

In order to solve the *Dynamic Cooperative Cleaners* problem we use the same cleaning protocol which was proposed for the static variant of the problem, as presented in the previous chapter, namely – the **SWEEP** protocol (see the complete definition of the protocol either in the previous chapter, or in Appendix). However, there are several implications to the fact that the contamination is now assumed to be expanding over time. First, the analysis of the protocol's performance becomes somewhat more complicated (moreover, the protocol's correctness itself must be redemonstrated). Second, a group of k agents can no longer be assumed to clean a contaminated region given enough time. Rather, the group of agents must be large enough, in order to guarantee a successful completion of the task. Finally, agents acting according to the **SWEEP** protocol, are expected to face a scenario which did not take place when the contaminated region was assumed to be static, namely — a spread of the contamination. Specifically, upon contamination spread, agents might find themselves located in tiles which are no longer boundary tiles. Once this happens, an agent will move towards the boundary tile which implements the continuation of his planned traversal along the region (this is executed through the movement to the *rightmost* neighbor, as defined in Definition 7).

Additional technical mechanisms which were discussed in the previous chapter (such as the *Pivot Point, Signaling, Connectivity Preservation, Agents' Synchronization, Clustering Problem* and *Mission Termination*) remain unaffected by the dynamic nature of the problem. For the purpose of making this chapter self-contained, they are redefined in Appendix.

4 Recursive Upper Bound

Due to the preservation of the *critical points*, such points can be visited many times by the agents before being cleaned. Furthermore, due to the dynamic nature of the problem, the shape of the contaminated region can change dramatically during the cleaning process.

Lemma 1 *The cardinality of the region's circumference always decreases after being traversed by an agent applying the* **SWEEP** *protocol (assuming no spread occurs), namely:*

$$|\partial F'| \leq |\partial F| - 8$$

Proof For any region F such that the number of tiles in F' is at least 2, note that a traversal around F is a closed, simple, rectilinear polygon. $\partial F'$ was obtained after deleting all the *non-critical* points of ∂F. Traversing such a polygon, an agent either goes straight, makes an internal turn ("left" turn, if we assume a clockwise movement), or makes an external turn ("right" turn). Suppose without loss of generality that an agent is moving up and making an internal turn left. Assume that there are no critical points. The path around this turn, along the newly created boundary tiles is now longer by two, as it contains an additional movement up, and another one towards the left. Similarly, an external turn, creates a new path which is shorter by two tile (as a single horizontal movement and a single vertical movement are no longer necessary). Since ∂F is a simple rectilinear polygon, it always has four "right" turns more than "left" ones.[1] When *critical* points are met, they are not being cleaned, which means that they exist in both ∂F and $\partial F'$. Re-visiting these points, however, does not change the overall size of the set of tiles (see an example in Fig. 6).

This proof, though, does not always hold for regions that, after being traversed once, produce either $\emptyset$ or a single tile. The reason is that the proof of the Lemma that appears in [39] relies on the fact that for each "turn" in F, there is a corresponding "turn" in F'. However, the concept of "turn" requires the existence of a possible movement of an agent from one tile to another time. This obviously cannot be done for $F' = \emptyset$ or for $F' = \{v\}$. For such regions it can easily be seen the Lemma holds, as $|\partial\emptyset| = 0$ and as $|\partial\{v\}| = 1$.

Lemma 2 *Every time a region is traversed by an agent using the* **SWEEP** *protocol, its width is decreased by at least one (assuming no spread occurs), namely:*

$$W(F'_t) \leq W(F_t) - 1$$

Proof By the definition of the depth of a tile, it is the shortest path to a *non-critical* tile in ∂F. According to the **SWEEP** protocol, after an agent had traversed F, all

---

[1]This is a simple consequence of the "rotation index" Theorem (see e.g. [21] p. 396): If $\alpha : [0, 1] \to R^2$ is a plane, regular, simple, closed curve, then $\int_0^1 k(s)ds = 2\pi$, where $k(s)$ is the curvature of $\alpha(s)$ and the curve is traversed in the positive direction.

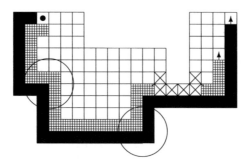

Fig. 6 An illustration of Lemma 1 and of the **SWEEP** protocol. The *black dot* denotes the starting point of the agents cleaning F (which is the entire region presented). The X's mark the *critical points* which are not cleaned. The *black tiles* are some of the tiles of ∂F, cleaned by the first agent. The second layer of marked tiles represent some of the tiles of $\partial F'$, cleaned by the second agent. The effect of corners on the traversal cardinality is either an increase (marked by the *left circle*) or a decrease (marked by the *right circle*)

of the *non-critical points* in ∂F were cleaned. This is true as the agent is instructed to clean all the *non-critical points* it is going through. The only exception are tiles which an agent does not clean upon exit, as there are other agents currently located in it. However, in this case, this tile will be cleaned by the last agent leaving it, at the following time steps. Therefore, after an agent traverses the region, the depth of every internal tile was decreased by (at least) one, meaning that the total width of the region was decreased by (at least) one. As a result, as all agents are identical, when k agents complete a traversal of F, the width of F is decreased by at least k.

Lemma 3 *The time it takes an agent which uses the* **SWEEP** *protocol to move along a path of length c_t (including delays caused by other agents located in the same tiles) is at most $4 \cdot c_t$.*

Proof According to the problem's specifications, each agent moves along the dirty region F at a pace of one tile per time step. The only exception may occur when several agents are delayed after entering the same tile.

According to the **SWEEP** protocol, although several agents can enter the same tile at the same time step, agents located in the same tile can leave it at the same time step only if they do so towards different directions. As a result, there arises the question whether more and more agents can enter a certain tile without leaving it, causing a decrease in the swarm's performance.

Let v be an empty tile. Let us assume without loss of generality that at some time step t, 4 agents enter v (this is of course the maximal number of agents which can enter v at the same time step, due to the invariant stating that no two agents can exit u to v at the same time step, for u, some neighbor of v, and since v has at most 4 neighbors). Thus, in time step $t - 1$ v had exactly 4 neighbors in ∂F. Let us assume that in time step $t + 1$ v still has 4 neighbors in ∂F. Thus, all the 4 agents which entered v intends to move to 4 different directions (since according to the **SWEEP** protocol, each of them has different *rightmost* neighbor). Thus, none of them will be delayed. Alternatively, let us assume that at time step $t + 1$ v has less than 4 neighbors in ∂F. Let α denote the number of neighbors of v in ∂F. Then, since the 4 agents which entered v in time t did so from different directions, according to the **SWEEP** protocol, α of them will be able to leave v towards the α neighbors of v, and $(4 - \alpha)$ will stay in v. However, in this time step, only α new agents can enter v, since v has only α neighbors in ∂F. Thus, in time step $t + 2$ there are at most 4 agents in v. This is true of course for each time steps $T > t + 2$. Since the number of agents in each tile is at most 4, and since each tile with agents in it has at least one neighbor of ∂F (since the agents had to enter it from a neighbor in ∂F), then there are at least $\frac{k}{4}$ agents which are able to move. Thus, even if the agents collide with each other continuously, the time it takes k agents to traverse a region of circumference c_t is at most $4 \cdot k \cdot c_t$ (and the average traversal time of an agent, amortized for the entire group of agents, is at most 4 times its minimal traversal time).

Notice that this still holds for the rare occasions in which the *check near completion of mission* procedure detects that every non-critical perimeter tile of F_t contains at least two agents. In his case, the *resting* commands are indeed overruled, allowing temporarily more than 4 agents to be located into the same tile, however — as the internal prioritization of the agents is maintained, this is corrected at the following time steps, as the "redundant" agents will wait (sometimes for more than 4 time steps). Notice though that this "long resting" was already compensated by the "untimely movement" those agents had performed while entering this tile.

It should be stated that in reality, the agents' traversal time is only slightly more than c_t, although an analytic proof is yet to be found.

Lemma 4 c_t, the length of the circumference of F_t never exceeds twice its cardinality, namely:

$$c_t \leq 2 \cdot |\partial F_t| - 2$$

Proof ∂F_t is a connected graph and thus it has spanning trees. An algorithm for constructing a spanning tree for ∂F_t and a *Depth First Search* (DFS) scan of it was constructed, such that the path generated by using the DFS algorithm contains a traversal around F_t (meaning that the DFS scan generates a path which after having several of its tiles removed, equals a path which traverses F_t).

This recursive spanning tree construction algorithm appears in Fig. 7. The algorithm receives p_s, a *non-critical* tile as its starting point and an empty list of tiles, L.[2] The algorithm constructs a spanning tree of ∂F_t as well as a DFS scan for this tree, by adding tiles to the list, while optionally marking some of them as tiles which should later be deleted. After the deletion of these tiles the remaining tiles form a path which traverses F_t. An example of the above appears in Fig. 8.

A DFS scan of a tree can go through each edge at most twice. A tree of $|V|$ vertices contains exactly $(|V| - 1)$ edges. Thus, a DFS scan of a spanning tree of ∂F_t contains at most $2 \cdot (|\partial F_t| - 1)$ transitions of edges. Since there exists such a spanning tree that contains a 'tour' around F_t then: $c_t \leq 2 \cdot |\partial F_t| - 2$.

---

[2] The existence of a *non-critical* point is guaranteed since ∂F_t is a connected graph and thus has a spanning tree, in which at least two tiles has a degree of 1, which makes them *non-critical* tiles.

1: Add (x, y) to L and mark it as **OLD**
2: Let v denote the *rightmost* neighbor of (x, y)

3: **If** $v \equiv p_s$ then
4:     Delete all the marked tiles from L
5:     **STOP**
6: **End if**

7: **If** (v is marked as **OLD**) and (no circle was formed) then
8:     /* The traversal repeats this tile as well */
9:     Call **SPAN-DFS** for v
10:    Upon return — **STOP**
11: **End if**

12: **If** (v is marked as **OLD**) and (a circle was formed) then
13:    /* Continue, and clean the redundant tiles */
14:    Add to L the sequence of tiles from L, starting with (x, y)
       and backwards to v
15:    Mark these tiles (excluding (x, y)) to be deleted
16:    Call **SPAN-DFS** for v
17:    Upon return — **STOP**
18: **End if**

19: /* The rightmost neighbor was not OLD */
20: Call **SPAN-DFS** for v
21: Upon return — **STOP**

Fig. 7 The recursive spanning tree construction algorithm. The term *rightmost* has the same meaning as in the **SWEEP** protocol. By checking L we can find out whether continuing to an **OLD** neighbor is a "clean return" (e.g. (A,B,C,B)) or whether it completes a circle (e.g. (A,B,C,A))

An important feature of a region F_t is its size — a bound over the contaminated region's size will later be used for producing an upper bound over the cleaning time of the agents. The following Lemma presents an upper bound over the size of the contaminated region at time t.

Lemma 5 *At any given time, the size of the contaminated region (i.e. $S_t = |F_t|$) can be bounded as follows:*

$$S_t \leq S_0 + \eta \cdot c_0 + 2(\eta^2 + \eta) \quad where \quad \eta \triangleq \left\lfloor \frac{t}{d} \right\rfloor$$

Proof The only time clean tiles can become contaminated is when the contamination spreads. When producing an upper bound for S_t, we can therefore disregard the cleaning done by the agents, as it can never result in the contamination of clean tiles. The maximal number of new contaminated tiles is achieved when the already

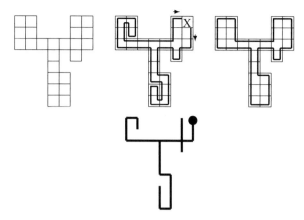

Fig. 8 An example of the spanning tree protocol. The *left chart* represents the dirty region F_t. The *middle chart* shows the DFS scan of ∂F_t according to the **SPAN-DFS** protocol, when the big X marks p_s, the *non-critical* starting point. The *right chart* shows the traversal path which is contained in this DFS scan. The drawing on the *bottom* shows the spanning tree that is created by the protocol and is searched by the DFS algorithm

contaminated tiles have the maximal boundary area possible. It is easy to see that this happens when the tiles are arranged in the form of a straight line. Since F_t is connected it has at least $S_t - 1$ edges, meaning that the sum of the degrees of the tiles is at least $2(S_t - 1)$. Thus, the maximal boundary area is at most $4S_t - 2(S_t - 1)$ (four possible neighbors per contaminated tile minus the edges already connecting the contaminated tiles). In a straight line all tiles but two have two clean neighbors while two tiles have three clean neighbors. This sums up to $2(S_t - 2) + 6$ which is exactly the maximal boundary area possible.

In such a case, when the contamination spreads for the i-th time, the area of the region is increased by $(c_0 + 2 + 2\iota_i)$ when ι_i is the i-th element of the series $(1, 3, 5, 7, \ldots)$. Let S_{s_i} denote the area of F_t after the i-th spread. Then for any region F_t, this is bounded as follows:

$$S_{s_i} \le S_0 + i(c_0 + 2) + 2\big(1 + 3 + 5 + \ldots + (2i - 1)\big)$$

The last can be simplified to:

$$S_{s_i} \le S_0 + i \cdot c_0 + 2(i^2 + i) \tag{1}$$

Since a spread occurs every d time steps, t can be written in a form of: $t = i \cdot d + \alpha$ where i is the number of spreads and α — the remainder. Since we are only interested in the number of spreads (i.e. i), we can use:

$$i = \left\lfloor \frac{t}{d} \right\rfloor$$

By applying this to (1) we get the requested result.

It is obvious that the number of tiles in the boundary of F_t is equal or smaller than its total size, namely:

$$\partial F_t \leq S_t$$

After combining this with Lemmas 4 and 5 and with the "intra-spreads $|\partial F_t|$ preservation" property of Lemma 1, we produce the following bound for c_t.

Lemma 6 *At any given time, the circumference of the contaminated region F_t is bounded as follows:*

$$c_{t+1} \leq 2\left[\Lambda_t + \eta \cdot c_0 + 2(\eta^2 + \eta) - 1\right]$$

where

$$\eta \triangleq \left\lfloor \frac{t}{d} \right\rfloor \quad and \quad \Lambda_t = \begin{cases} |\partial F_0|, & t \leq d \\ S_0, & t > d \end{cases}$$

Proof According to Lemma 4, between two spreads, c_t is bounded by twice its original value (from just after the last spread). After a spread occurs, the value of c_t is being reset, according to the bound for the total size of F_t (Lemma 5).

It is easy to see that the width of F_0 at any time is bounded by its initial width plus the number of contamination spreads. This notion is expressed in the following Lemma.

Lemma 7

$$W('F_t) \leq W(F_t) + 1$$

$$W(F_{t+1}) \leq W(F_0) + \eta \quad where \quad \eta \triangleq \left\lfloor \frac{t}{d} \right\rfloor$$

With $W_{REMOVED}(t)$ denoting the decrease in F's width due to the agents cleaning activity until time t, we can now have the following:

Lemma 8

$$If \quad \left\lfloor \sum_{i=1}^{t} \frac{k}{4 \cdot c_i} \right\rfloor \geq \overline{W}(F_{t+1}) \quad then \quad W_{REMOVED}(t) \geq \overline{W}(F_{t+1})$$

where $\overline{W}(F_{t+1})$ denotes the width of F_0 at time $t + 1$, assuming that no cleaning took place by the agents thus far (namely, the width predicted by Lemma 7).

Proof $W_{REMOVED}(t)$ can be defined as the number of completed traversals around the contaminated region, multiplied by the number of the agents k. Note that at time step t, each agent completes $\frac{1}{c_t}$ of the circumference. Since we know the value of c_t for every t, and since we know that the time it takes an agent to move along a path is at most 4 times the length of this path (see Lemma 3), then the decrease in the original width of the region caused by the cleaning process of the agents is bounded as described.

Note that as the expression which appears in the Lemma does not distinguish between a multitude of agents working for a short while, and a single agent which is working for a long period of time, it may not hold for smaller values of t (in which the accumulated partial peelings is smaller than $\overline{W}(F_{t+1})$). Therefore, while writing $W_{REMOVED}(t) \geq \left\lfloor \sum_{i=1}^{t} \frac{k}{4 \cdot c_i} \right\rfloor$ might not always hold, the expression as appears above does.

Note that this holds also when taking into account the effect the spreading contamination has on the region's width. Let $p = (v_1, v_2, \ldots, v_n)$ denote a "tour", traversing a *contaminated* region F, and let us assume that a spread occurs at time $t = t_s$. Let $p = p_1 \cdot p_2$, and let p_1 denote the part of the 'tour' which took place before the contamination spread.

The width of the region F is defined as the depth of the deepest tile in F. Let us assume without loss of generality that there is only one tile of maximal depth, u (if there is more than a single deepest tile, the same observation can be applied to all deepest tiles, separately). Therefore, $W(F_t)$ is the length of the shortest *path* (or paths) from u to a *non-critical* tile in ∂F_t. Let us assume without loss of generality that there is only one such shortest path, from u to a *non-critical* boundary tile u_{END} (if there are more than a single shortest path, the same observation can be applied to all, separately). u_{END} is a boundary tile, and therefore must be included at least once in p (either in p_1 or in p_2, or in both). If $(u_{END} \in p_1 \ \wedge \ u_{END} \in p_2)$ then u_{END} is a *critical point*(and we already know this is not the case).

If $(u_{END} \in p_1)$ then it was already cleaned prior to t_s, decreasing the width of the region by 1. After the contamination spread, u_{END} might become *contaminated* once again, restoring the region's width to its original value.

If $\neg(u_{END} \in p_1)$ it therefore means that $(u_{END} \in p_2)$. As u_{END} is a boundary tile of F it has at least one $8 - Neighbor$ which is clean prior to t_s. In order for the contamination spread to increase the width of F, it means that all the $8 - Neighbors$ of u_{END} which were clean have become contaminated after the spread. However, as those neighbors are perimeter tiles, they will be visited by the agent. As at least one of them can be cleaned (there does not exist a tile whose $8 - Neighbors$ are all critical points), u_{END} will become a perimeter tile again, restoring the original value of the width of F.

Spreads can naturally occur more than once in the middle of a 'tour'. In this case, the same principal can be applied recursively. An example of the above appears in Fig. 9.

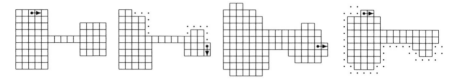

Fig. 9 An example of an "interrupting spread" — a spread that occurs in a middle of a 'tour'. The *first chart* on the *left* is the original contaminated region, whose width is 3 (the *black circle* is the agent and the *arrow* marks its next movement). The *second chart* is the region just before the spread (the *dots* are the cleaned tiles). The *third chart* is the region just after the spread (notice that the width is now 4 and that the 'tour' was changed). The *fourth chart* is the region after the agent completes its (new) 'tour'. Notice that the width is 3 again

An upper bound over the cleaning time of a contaminated region F_0 can now be constructed, as follows:

Theorem 1 *Assume that k agents start cleaning a simply connected region F_0 at some boundary point p_0 and work according to the* **SWEEP** *protocol, and denote by $t_{success}(k)$ the time needed for this group to clean F_0. Then it holds that:*

1. *If $(t = \frac{8(|\partial F_0|-1)\cdot(W(F_0)+k)}{k} + 2k)$ is not greater than d, then $t_{SUCCESS} = \lceil t \rceil$. This also holds for static environments, since in such cases $d \to \infty$.*
2. *Otherwise $(t > d)$: $t_{SUCCESS}$ is the minimal (integer) t for which:*

$$\sum_{i=d+1}^{t} \frac{1}{S_0 - 1 + 2\lfloor \frac{i}{d} \rfloor^2 + (c_0 + 2)\lfloor \frac{i}{d} \rfloor} \geq \gamma + \frac{8}{k} \cdot \left\lfloor \frac{t}{d} \right\rfloor$$

where

$$\gamma \triangleq \frac{8(k + W(F_0))}{k} - \frac{d - 2k}{|\partial F_0| - 1}$$

Proof In order for F_0 to be cleaned, there should exist a time $t_{SUCCESS}$ in which the width of the region will be 0. Remembering that $W(F_t)$ is monotonically increasing and disregards the cleaning performed by the agents, and that $W_{REMOVED}(t)$ denotes the decrease in the width of F_0 due to the cleaning of the agents, we are interested in:

$$W(F_{t_{SUCCESS}}) + k \leq W_{REMOVED}(t_{SUCCESS})$$

The purpose of the "$+k$" is to guarantee the cleaning of the 'skeleton' that remains when the width of the region decreases to zero. In such a time, the agents function as a single agent, and we must guarantee that an additional traversal will be performed. We now apply Lemmas 7 and 8.

Note that either $t_{SUCCESS} \leq d$ or $t_{SUCCESS} > d$. We first examine the case of $t_{SUCCESS} \leq d$. Regarding $W(F_t)$, c_t and $W_{REMOVED}(t)$ this means that:

$$\forall i \leq t , \ W(F_i) \leq W(F_0)$$

$$\forall i \leq t , \ c_{i+1} \leq 2(|\partial F_0| - 1)$$

$$W_{REMOVED}(t) \geq \left\lfloor \sum_{i=1}^{t} \frac{k}{4 \cdot c_i} \right\rfloor \geq \left\lfloor \frac{k \cdot t}{8|\partial F_0| - 8} \right\rfloor$$

Thus, we are interested in:

$$\left\lfloor \frac{k \cdot t_{SUCCESS}}{8(|\partial F_0| - 1)} \right\rfloor \geq W(F_0) + k$$

Since releasing the agents requires an additional $2k$ time steps, the final bound for this case is:

$$t_{SUCCESS} \geq \frac{8(|\partial F_0| - 1) \cdot (W(F_0) + k)}{k} + 2k$$

If the $t_{SUCCESS}$ received is greater than d, we continue to the next step. Lemma 6 provides us with c_t. As for $W_{REMOVED}(t_{SUCCESS})$, remembering that the actual cleaning begins at $t = 2 \cdot k$ (the time it takes to release the agents) we get:

$$W_{REMOVED}(t_{SUCCESS}) \geq$$

$$\geq \left\lfloor \sum_{i=2k}^{t_{SUCCESS}} \frac{k}{4 \cdot c_i} \right\rfloor \geq \left\lfloor \sum_{i=2k}^{d} \frac{k}{4 \cdot c_i} \right\rfloor + \left\lfloor \sum_{i=d+1}^{t_{SUCCESS}} \frac{k}{4 \cdot c_i} \right\rfloor \geq$$

$$\geq \left\lfloor \frac{(d - 2k) \cdot k}{8(|\partial F_0| - 1)} \right\rfloor + \left\lfloor \sum_{i=d+1}^{t_{SUCCESS}} \frac{k}{8 \cdot (S_0 + \lfloor \frac{i}{d} \rfloor \cdot c_0 + 2(\lfloor \frac{i}{d} \rfloor^2 + \lfloor \frac{i}{d} \rfloor) - 1)} \right\rfloor$$

As we already know that $W(F_{i+1}) \leq W(F_0) + \lfloor \frac{i}{d} \rfloor$ using the previous expression for $W_{REMOVED}$ yields:

$$\left\lfloor \sum_{i=2k}^{d} \frac{k}{4 \cdot c_i} \right\rfloor + \left\lfloor \sum_{i=d+1}^{t} \frac{k}{4 \cdot c_i} \right\rfloor \geq W(F_0) + k + \left\lfloor \frac{t}{d} \right\rfloor$$

$$\implies \left\lfloor \sum_{i=d+1}^{t} \frac{k}{4 \cdot c_i} \right\rfloor \geq W(F_0) + k + \left\lfloor \frac{t}{d} \right\rfloor - \frac{(d-2k) \cdot k}{8(|\partial F_0| - 1)}$$

$$\implies \sum_{i=d+1}^{t} \frac{1}{S_0 - 1 + 2\lfloor \frac{i}{d} \rfloor^2 + (c_0 + 2)\lfloor \frac{i}{d} \rfloor} \geq$$

$$\geq \left(\frac{8(k + W(F_0))}{k} - \frac{d-2k}{|\partial F_0| - 1} \right) + \frac{8}{k} \cdot \left\lfloor \frac{t}{d} \right\rfloor$$

In order to find such t which satisfies this expression, one should add elements to the sum iteratively, recalculating the value of the right expression. The minimal t for which the above holds is $t_{SUCCESS}$. Performing the above takes $O(t_{SUCCESS})$ time (since every iteration requires a single comparison).

Notice that the expression $t_{NoSpread}(k) \triangleq \frac{8(|\partial F_0|-1) \cdot (W(F_0)+k)}{k} + 2k$ is not monotonic. Moreover, we can see that:

$$\lim_{k \to \infty} t_{NoSpread}(k) = \infty$$

Note that given k agents, using merely a portion of them is naturally a valid strategy, should better results are to be obtained by such action. Therefore, we can state that:

Corollary 1 *Given a contaminated region F_0 and k cleaning agents using the* **SWEEP** *protocol. Let us define:*

$$\hat{k} \triangleq \sqrt{4 \cdot W(F_0) \cdot (|\partial F_0| - 1)}$$

and let $\hat{t}_{NoSpread}$ be defined as:

$$\min \left\{ \frac{8(|\partial F_0| - 1) \cdot (W(F_0) + \lceil \hat{k} \rceil)}{\lceil \hat{k} \rceil} + 2\lceil \hat{k} \rceil, \frac{8(|\partial F_0| - 1) \cdot (W(F_0) + \lfloor \hat{k} \rfloor)}{\lfloor \hat{k} \rfloor} + 2\lfloor \hat{k} \rfloor \right\}$$

If $\hat{k} < k$ and $\hat{t}_{NoSpread} \leq d$ then the completion of the cleaning mission before a contamination spread is guaranteed, and $t_{success} = \lceil \hat{t}_{NoSpread} \rceil$.

Proof We are interested in the "effective" number of agents $\hat{k}$, which minimizes the value of $t_{NoSpread}$. Namely, a value $\hat{k}$ for which $\frac{dt_{NoSpread}}{dk}(\hat{k}) = 0$, provided that $\hat{k} \leq k$.

Theorem 1 is illustrated in Figs. 10 and 11.

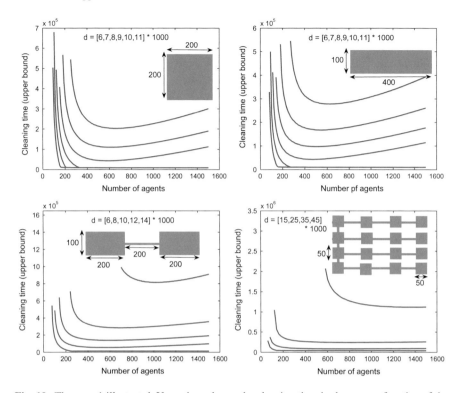

Fig. 10 Theorem 1 illustrated. Upper bound over the cleaning time is shown as a function of the number of cleaning agents using the **SWEEP** protocol. All four contaminated regions shown are of the same initial size

Fig. 11 Theorem 1 illustrated. Upper bound over the cleaning time is shown as a function of the number of cleaning agents using the **SWEEP** protocol, for $d = 10000$ and size of initial region varying between $S_0 = 10000$ and $S_0 = 90000$ (all other geometric properties remaining the same)

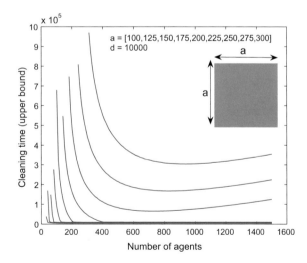

5   Direct Upper Bounds

The bound in Theorem 1 contains a recursive expression and requires an iterative process in order to calculate the minimal time $t_{SUCCESS}$ in which it is guaranteed that the agents will complete the cleaning mission. The following Theorem presents an upper bound for the cleaning time given in a form of a direct expression, whose solution is the requested value for $t_{SUCCESS}$.

Theorem 2 *If* $(t = \frac{8(|\partial F_0|-1)\cdot(W(F_0)+k)}{k} + 2k)$ *is not greater than* d, *then* $t_{SUCCESS} = \lceil t \rceil$. *This also holds for static environments, since in such cases* $d \to \infty$.

  Otherwise $(t > d)$: *find the minimal* μ *for which:*

$$\psi\left(\mu + \frac{c_0 + 2 - \gamma_2}{4}\right) - \psi\left(\mu + \frac{c_0 + 2 + \gamma_2}{4}\right) + \gamma_1 - \frac{\gamma_2 \cdot \gamma}{d} - \frac{8 \cdot \gamma_2}{d \cdot k} \cdot \mu \geq 0$$

where

$$\gamma_1 \triangleq \psi\left(1 + \frac{c_0 + 2 + \gamma_2}{4}\right) - \psi\left(1 + \frac{c_0 + 2 - \gamma_2}{4}\right)$$

and

$$\gamma_2 \triangleq \sqrt{(c_0 + 2)^2 - 8 \cdot (S_0 - 1)} \quad and \quad \gamma \triangleq \frac{8(k + W(F_0))}{k} - \frac{d - 2k}{|\partial F_0| - 1}$$

where $\psi()$ *is the* Digamma *function (studied in [1]) — the logarithmic derivative of the* Gamma *function, defined as:*

$$\psi(x) = \frac{d}{dx} \ln \Gamma(x) = \frac{\Gamma'(x)}{\Gamma(x)}$$

or as:

$$\psi(x) = \int_0^\infty \left(\frac{e^{-t}}{t} - \frac{e^{-xt}}{1 - e^{-t}}\right) dt$$

Then $t_{SUCCESS} = \lceil \mu \cdot d \rceil$.

Proof The first part of the Theorem is equivalent to the first part of Theorem 1.

  For the second part, recall that in the second part of Theorem 1 we were interested in finding the minimal t for which the following inequality holds:

$$\sum_{i=d+1}^{t} \frac{1}{S_0 - 1 + 2\lfloor \frac{i}{d} \rfloor^2 + (c_0 + 2)\lfloor \frac{i}{d} \rfloor} \geq \gamma + \frac{8}{k} \cdot \left\lfloor \frac{t}{d} \right\rfloor$$

where

$$\gamma \triangleq \frac{8(k + W(F_0))}{k} - \frac{d - 2k}{|\partial F_0| - 1}$$

Let us denote $\mu \triangleq \lfloor \frac{t}{d} \rfloor$. Therefore we can now search for the minimal μ for which:

$$\sum_{i=1}^{\mu} \frac{d}{S_0 - 1 + 2 \cdot i^2 + (c_0 + 2) \cdot i} \geq \gamma + \frac{8}{k} \cdot \mu$$

After extracting d we get:

$$d \cdot \sum_{i=1}^{\mu} \frac{1}{S_0 - 1 + 2 \cdot i^2 + (c_0 + 2) \cdot i} \geq \gamma + \frac{8}{k} \cdot \mu$$

Observe that the left side of the inequality is in the form of:

$$d \cdot \sum_{i=1}^{\mu} \frac{1}{A + B \cdot i + C \cdot i^2} \tag{2}$$

where $A = (S_0 - 1) \quad B = (c_0 + 2) \quad C = 2$.
This expression can be integrated[3] into the form of:

$$d \cdot \left. \frac{\psi(\frac{2 \cdot C \cdot x + B - \sqrt{B^2 - 4 \cdot C \cdot A}}{2 \cdot C}) - \psi(\frac{2 \cdot C \cdot x + B + \sqrt{B^2 - 4 \cdot C \cdot A}}{2 \cdot C})}{\sqrt{B^2 - 4 \cdot C \cdot A}} \right|_1^{\mu} \tag{3}$$

After assigning the proper values of A, B and C, we receive the following formula, where as the minimal μ which satisfies it, will be denoted as $\mu_{SUCCESS}$:

$$\frac{d}{\gamma_2} \cdot \left(\psi(\mu + \frac{c_0 + 2 - \gamma_2}{4}) - \psi(\mu + \frac{c_0 + 2 + \gamma_2}{4}) \right) + \frac{d}{\gamma_2} \cdot \gamma_1 \geq \gamma + \frac{8}{k} \cdot \mu$$

where:

$$\gamma_1 \triangleq \psi\left(1 + \frac{c_0 + 2 + \gamma_2}{4}\right) - \psi\left(1 + \frac{c_0 + 2 - \gamma_2}{4}\right)$$

and:

$$\gamma_2 \triangleq \sqrt{(c_0 + 2)^2 - 8 \cdot (S_0 - 1)}$$

By rewriting the above, we get the following formula:

$$\psi\left(\mu + \frac{c_0 + 2 - \gamma_2}{4}\right) - \psi\left(\mu + \frac{c_0 + 2 + \gamma_2}{4}\right) + \gamma_1 - \frac{\gamma_2 \cdot \gamma}{d} - \frac{8 \cdot \gamma_2}{d \cdot k} \cdot \mu \geq 0$$

where:

$$\gamma_1 \triangleq \left(\psi(1 + \frac{c_0 + 2 + \gamma_2}{4}) - \psi(1 + \frac{c_0 + 2 - \gamma_2}{4}) \right)$$

---

[3]The integral was produced using MATLAB [38].

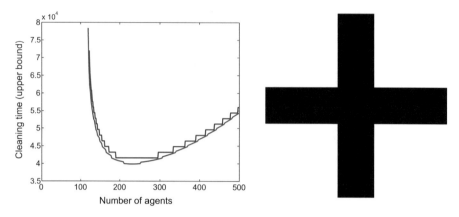

Fig. 12 An example of Theorem 2. The lower curve represents a bound over the cleaning time, as produced by Theorem 1, for the cross region which appears on the right (for which $S_0 = 2960$, $c_0 = 326$, $d = 1600$). The upper curve is produced by using Theorem 2

and:

$$\gamma_2 \triangleq \sqrt{(c_0 + 2)^2 - 8 \cdot (S_0 - 1)}$$

and:

$$\gamma \triangleq \left(\frac{8(k + W(F_0))}{k} - \frac{d - 2k}{|\partial F_0| - 1} \right)$$

Since $\mu = \lfloor \frac{t}{d} \rfloor$ then $\mu > \frac{t}{d} + 1$. Thus, $t_{SUCCESS} < (\mu_{SUCCESS} - 1) \cdot d$.

Figure 12 shows an example of applying Theorem 2 for a cross shaped region.

6 Simplifying Theorem 2

Although Theorem 2 presents a direct way of calculating an upper bound over the cleaning time of k agents using the **SWEEP** protocol, this expression is still implicit. Theorem 3 presents an explicit approximations of the result of Theorem 2.

Theorem 3 *Let F_0 be a contaminated region for which Theorem 2 predicts that the contamination will spread before being cleaned. Then let $\mu_{SUCCESS} \triangleq \min \{x \in \{\mu_1, \mu_2\} | x > 0\}$ where μ_1 and μ_2 are defined as:*

$$\mu_1, \mu_2 = \frac{-(A_1 A_3 - A_4) \pm \sqrt{(A_1 A_3 - A_4)^2 - 4A_3(A_2 - A_1 - A_1 A_4)}}{2A_3}$$

where:

$$A_1 = \frac{c_0 + 2 - \gamma_2}{4}$$

$$A_2 = \frac{c_0 + 2 + \gamma_2}{4}$$

$$A_3 = \frac{8 \cdot \gamma_2}{d \cdot k}$$

$$A_4 = \gamma_1 - \frac{\gamma_2 \cdot \gamma}{d}$$

and where:

$$\gamma_1 \triangleq \psi\left(1 + \frac{c_0 + 2 + \gamma_2}{4}\right) - \psi\left(1 + \frac{c_0 + 2 - \gamma_2}{4}\right)$$

and:

$$\gamma_2 \triangleq \sqrt{(c_0 + 2)^2 - 8 \cdot (S_0 - 1)}$$

and:

$$\gamma \triangleq \frac{8(k + W(F_0))}{k} - \frac{d - 2k}{|\partial F_0| - 1}$$

If such $\mu_{SUCCESS}$ exists then:

$$t_{SUCCESS} = (\mu_{SUCCESS} - 1) \cdot d$$

Proof From Theorem 2 we know that if $\mu_{SUCCESS}$ is the minimal μ for which:

$$\psi\left(\mu + \frac{c_0 + 2 - \gamma_2}{4}\right) - \psi\left(\mu + \frac{c_0 + 2 + \gamma_2}{4}\right) + \gamma_1 - \frac{\gamma_2 \cdot \gamma}{d} - \frac{8 \cdot \gamma_2}{d \cdot k} \cdot \mu \geq 0$$

(for some constant γ, γ_1 and γ_2) then:

$$t_{SUCCESS} = (\mu_{SUCCESS} - 1) \cdot d$$

Note that this expression is in the form of:

$$\psi\left(\mu + A_1\right) - \psi\left(\mu + A_2\right) - A_3 \cdot \mu + A_4 \geq 0 \qquad (4)$$

As to the *digamma* function, we know that (Eq. 6.3.5 in [1]):

$$\psi(z + 1) = \psi(z) + \frac{1}{z}$$

Thus, we can see that:

$$\frac{A_1 - A_2}{\mu + A_1} \leq \psi\left(\mu + A_1\right) - \psi\left(\mu + A_2\right) \leq \frac{A_1 - A_2}{\mu + A_2} \tag{5}$$

Combining (4) and (5) we can see that we must find the minimal μ for which:

$$\frac{A_1 - A_2}{\mu + A_1} \geq A_3 \cdot \mu - A_4 \tag{6}$$

Meaning that we must find the minimal μ for which:

$$A_3 \cdot \mu^2 + (A_1 A_3 - A_4) \cdot \mu + (A_2 - A_1 - A_1 A_4) \leq 0 \tag{7}$$

Note that $A_3 > 0$ and that $\mu \in (0..\infty)$. Thus, both roots are of the same sign (where for negative roots there is no valid solution, and for positive roots the minimal of them is the solution). Roots of opposites signs are not possible, since it implies that the solution is $\mu = 0$, which in turn is impossible as $F_0 \neq \emptyset$, and as at least one contamination spread took place. μ_1 and μ_2 are defined as:

$$\mu_1, \mu_2 = \frac{-(A_1 A_3 - A_4) \pm \sqrt{(A_1 A_3 - A_4)^2 - 4A_3(A_2 - A_1 - A_1 A_4)}}{2A_3}$$

After some arithmetics, and by assigning the values of A_1, A_2, A_3 and A_4 we get the requested result.

7 An Upper Bound for k

Using the previous upper bound for the time it takes k agents to clean an initial contaminated region of known geometric properties, a similar bound for the number of agents required to guarantee a successful cleaning of the region (in a finite amount of time) can be produced.

Theorem 4 *Given a contaminated region of properties S_0, $|\partial F_0|$ and $W(F_0)$, spreading every d time steps, then in order to guarantee a successful cleaning of the region before the contamination is able to spread even once, the minimal number of cleaning agents required is upper bounded as follows:*

$$(k_1 \leq k \leq k_2) \wedge (k > 0)$$

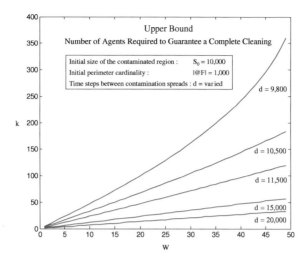

Fig. 13 Theorem 4 illustrated. The graph presents an upper bound over the number of agents needed to guarantee a successful cleaning of a region of initial size of 10,000 tiles and a perimeter cardinality of 1,000 tiles, as a function of the width of the region (namely — its "bulkiness"). Notice how the effect this geometric feature has on the number of agents required increases dramatically as the time intervals between contamination spreads decreases

where:

$$k_{1,2} = 2\left(-|\partial F_0| + 1 + \frac{d}{8}\right) \pm 2\sqrt{\left(|\partial F_0| - 1 - \frac{d}{8}\right)^2 - (|\partial F_0| - 1)W(F_0)}$$

Proof Let us use the first part of Theorem 2:

$$\frac{8(|\partial F_0| - 1) \cdot (W(F_0) + k)}{k} + 2k \leq d$$

After some arithmetics we see that:

$$2k^2 + (8|\partial F_0| - 8 - d)k + (8|\partial F_0| - 8)W(F_0) \leq 0$$

After extracting the value of k we get:

$$k_{1,2} = \frac{-(8|\partial F_0| - 8 - d) \pm \sqrt{(8|\partial F_0| - 8 - d)^2 - 8(8|\partial F_0| - 8)W(F_0)}}{4}$$

and the rest is implied (Figs. 13 and 14).

Fig. 14 Theorem 4 illustrated. Observe how as $\frac{S_0}{\partial F_0}$ increases (namely, the initial region is "bulkier"), it becomes easier to guarantee the proper cleaning of the region with a smaller number of agents. The width of the initial region was selected to reflect a reasonable proportion considering the changing cardinality value of the perimeter ∂F_0

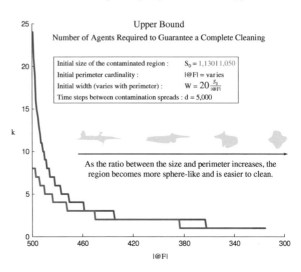

8   Discussion and Conclusion

This work described the *Dynamic Cooperative Cleaners* problem, where a swarm of simple and limited drones must clean a dynamic "contaminated" region spreading due to contamination of neighbors.

This problem has several interesting applications, some of which are for example an autonomous patrolling of a pre-defined area using a swarm of collaborative drones [8, 11, 16], or an implementation of a distributed anti-virus system for computer networks [15, 26]. Additional applications are distributed search engines [19], as well as the coordination of fire-fighting units (see [20] for details). The problem is also related to the geometric problem of pocket-machining [24]. An interesting problem of cleaning and maintenance of a system of pipes by an autonomous agent is discussed in [31] and a multi-agent version can be of much interest. The importance of cleaning hazardous waste by agents is described in [23].

SWEEP, a collaborative cleaning protocol to be implemented by each drone was presented and its performance was analyzed and several analytic upper bounds on its completion time were discussed.

As mentioned previously, our cooperative algorithms approach can be considered as a case of social behavior in the sense of [36], where one induces multi-agent cooperation by forcing the agents to obey some simple "social behavior" guidelines.

As in the case of the static variant of the problem, the question of how fast can k drones scan a pre-defined expanding region, is *NP-hard* (for any number of drones, as this can be reduced to finding a Hamilton path in a non-simple grid-graph [25]). Complexity analysis for this can be found in [5, 6], various effects on the completion time as a function of the region's geometric properties are analyzed in [9, 12], and upper bounds are discussed in [14]. Stochastic variant of this problem, involving a probabilistic analysis of the search problem in domains that spread nondeterministically is discussed in [34].

Appendix – Definitions

This Appendix contains terms which were defined in previous chapters, as well as several algorithms and proofs, reiterated for the purpose of self-containment.

The SWEEP Cleaning Protocol

The **SWEEP** protocol is implemented by each agent a_i, located at time t at $\tau_i(t) = (x, y)$. We define below several terms we use while discussing the **SWEEP** protocol. We stress the fact that this is indeed a *myopic* protocol, relying on neighborhood information, a *senile* protocol (the memory needed is constant + one counter whose size is upper bounded by $O(\log k)$, where k is the number of agents) and relies on *implicit local communication* only.

Definition 5 Let ∂F denote the boundary of F. A tile is on the boundary if and only if at least one of its $8Neighbors$ is not in F, meaning:

$$\partial F = \{v \mid v \in F \wedge 8Neighbors(v) \cap (G \setminus F) \neq \emptyset\}$$

Definition 6 Let $\tilde{\tau}_i(t)$ denote the previous location of agent i with respect to $\tau_i(t)$, such that $\tilde{\tau}_i(t) \neq \tau_i(t)$, defined as:

$$\tilde{\tau}_i(t) \triangleq \tau_i(x) \ s.t. \ x = \max\{j \in \mathbb{N} \mid j < t \, and \, \tau_i(j) \neq \tau_i(t)\}$$

Definition 7 The term '*rightmost*' means:

Starting from $\tilde{\tau}_i(t)$ (namely, the previous boundary tile that agent a_i had been in) scan the four neighbors of $\tau_i(t)$ in a clockwise order until a boundary tile (excluding $\tilde{\tau}_i(t)$) is found. Sometimes, $\tilde{\tau}_i(t)$ might not be a boundary tile (this may occasionally happen after a contamination spread, for dirty regions of specific geometric features. In those cases, an agent may find itself in an internal point — a dirty tile which is not located on the boundary. From this tile the agent must move to the adjacent boundary tile, and resume its traversal along the region's boundary). In this case, if $\tau_i(t)$ is a boundary point, then starting from $\tilde{\tau}_i(t)$ scan the four neighbors of $\tau_i(t)$ in a clockwise order until the second boundary tile is found. In case $t = 0$ select the tile as instructed in Fig. 15.

The reference to *contamination spread* used in Definition 7 is given with consideration to the use of this term in the next chapters.

The additional information needed for the protocol and its sub-routines is contained in $\mathcal{M}_i$ and $Neighborhood(x, y)$.

A schematic flowchart of the protocol, describing its major components and procedures is presented in Fig. 16. The complete pseudo-code of the protocol and its

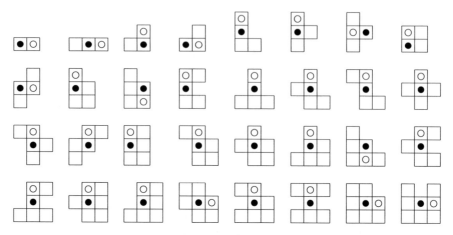

Fig. 15 When $t = 0$ the first movement of an agent located in (x, y) should be decided according to initial contamination status of the neighbors of (x, y), as appears in these charts — the agent's initial location is marked with a *filled circle* while the destination is marked with an *empty one*. All configurations which do not appear in these charts can be obtained by using rotations. This definition is needed in order to initialize the traversal behavior of the agents in the correct direction

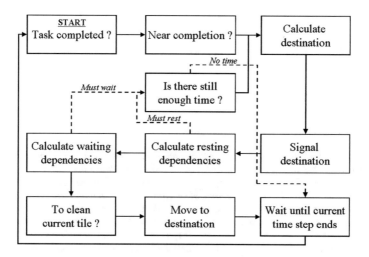

Fig. 16 A schematic flow chart of the **SWEEP** protocol. The *smooth lines* represent the basic flow of the protocol while the *dashed lines* represent cases in which the flow is interrupted. Such interruptions occur when an agent calculates that it must not move until other agents do so (either as a result of *waiting* or *resting* dependencies — see Lines 6 and 19 of **SWEEP** for more details)

sub-routines appears in Figs. 17 and 18. Upon initialization of the system, the *System Initialization* procedure is called (defined in Fig. 17). This procedure sets various initial values of the agents, and call the protocol's main procedure — *SWEEP* (defined in Fig. 18). This procedure in turn, uses various sub-routines and functions, all defined in Fig. 17.

```
 1:  System Initialization
 2:       Arbitrarily choose a pivot point p₀ in ∂F₀, and mark it as critical point
 3:       Place all the agents on p₀
 4:       For (i = 1; i ≤ k; i + +) do
 5:            Call Agent Reset for agent i
 6:            Call SWEEP for agent i
 7:            Wait two time steps
 8:       End for
 9:  End procedure

10:  Agent Reset
11:       resting ← false
12:       dest ← null /* destination */
13:       near completion ← false
14:       saturated perimeter ← false
15:       waiting ← ∅
16:  End procedure

17:  Priority
18:       /* Assuming the agent moved from (x₀, y₀) to (x₁, y₁) */
19:       priority ← 2(x₁ − x₀) + (y₁ − y₀)
20:  End procedure

21:  Check Completion of Mission
22:       If ((x, y) = p₀) and (x, y) has no dirty neighbors then
23:            If (x, y) is dirty then
24:                 Clean (x, y)
25:            STOP
26:  End procedure

27:  Check "Near Completion" of Mission
28:       /* Cases where every tile in Fₜ contains at least a single agent */
29:       near completion ← false
30:       If each of the dirty neighbors of (x, y) contains at least one agent then
31:            near completion ← true
32:       If each of the dirty neighbors of (x, y) has near completion = true then
33:            Clean (x, y) and STOP
34:       /* Cases where every non-critical tile in ∂Fₜ contains at least 2 agents */
35:       saturated perimeter ← false
36:       If ((x, y) ∈ ∂Fₜ) and both (x, y) and all of its non-critical neighbors
            in ∂Fₜ contain at least two agents then
37:            saturated perimeter ← true
38:       If ((x, y) ∈ ∂Fₜ) and both (x, y) and all of its neighbors in ∂Fₜ has
            saturated perimeter = true then
39:            Ignore resting commands for this time step
40:  End procedure
```

Fig. 17 The first part of the **SWEEP** cleaning protocol

Pivot Point

The initial location of the agents (the *pivot point*, denoted as p_0) is artificially set to be critical during the execution, hence it is also guaranteed to be the last point cleaned. Completion of the mission can therefore be concluded from the fact that all (working) agents are back at p_0 with no contaminated neighbors to go to, thereby reporting on completion of their individual missions. Note that this artificial preservation of the

```
 1:  SWEEP Protocol /* Controls agent i after Agent Reset */
 2:      Check Completion of Mission
 3:      Check "Near Completion" of Mission
 4:      dest ← rightmost neighbor of (x,y) /* Calculate destination */
 5:      destination signal bits ← dest /* Signaling the desired destination */
 6:      /* Calculate resting dependencies (solves agents' clustering problem) */
 7:      Let all agents in (x,y) except agent i be divided to the following groups :
 8:          A₁ : Agents signaling towards any direction different than dest
 9:          A₂ : Agents signaling towards dest which entered (x,y) before agent i
10:          A₃ : Agents signaling towards dest which entered (x,y) after agent i
11:          A₄ : Agents signaling towards dest which entered (x,y) with agent i
12:      Let group A₄ be divided into the following two groups :
13:          A₄ₐ : Agents with lower priority than this of agent i
14:          A₄ᵦ : Agents with higher priority than this of agent i
15:      resting ← false
16:      If (A₂ ≠ ∅) or (A₄ᵦ ≠ ∅) then
17:          resting ← true
18:          If (current time-step 𝒯 did not end yet) then jump to 4 Else jump to 35
19:      waiting ← ∅ /* Waiting dependencies (agents synchronization) */
20:      Let active agent denote a non-resting agent which didn't move in 𝒯 yet
21:      If (x-1,y) ∈ Fₜ contains an active agent then waiting ← waiting ∪ {left}
22:      If (x,y-1) ∈ Fₜ contains an active agent then waiting ← waiting ∪ {down}
23:      If (x-1,y-1) ∈ Fₜ contains an active agent then waiting ← waiting ∪ {l-d}
24:      If (x+1,y-1) ∈ Fₜ contains an active agent then waiting ← waiting ∪ {r-d}
25:      If dest = right and (x+1,y) contains an active agent j, and destⱼ ≠ left, and
         there are no other agents delayed by agent i (i.e. (x-1,y) does not contain
         active agent l with destₗ =right and no active agents in (x,y+1),(x+1,y+1),
         (x-1,y+1), and (x+1,y) does not contain active agent n with destₙ = left),
         then (waiting ← waiting ∪ {right}) and (waitingⱼ ← waitingⱼ \ {left})
26:      If dest = up and (x,y+1) contains an active agent j, and destⱼ ≠ down, and
         there are no other agents delayed by agent i (i.e. (x,y-1) does not contain
         active agent l with destₗ =up and no active agents in (x+1,y),(x+1,y+1),
         (x-1,y+1), and (x,y+1) does not contain active agent n with destₙ = down),
         then (waiting ← waiting ∪ {up}) and (waitingⱼ ← waitingⱼ \ {down})
27:      If (waiting ≠ ∅) then
28:          If (𝒯 has not ended yet) then jump to 4 Else jump to 35
29:      /* Decide whether or not (x,y) should be cleaned */
30:      If ¬ ((x,y) ∈ ∂Fₜ) or ((x,y) ≡ p₀) or (x,y) has 2 dirty tiles in its 4Neighbors
         which are not connected via a path of dirty tiles from its 8Neighbors then
31:          (x,y) is an internal point or a critical point and should not be cleaned
32:      Else
33:          Clean (x,y) if and only if it does not still contain other agents
34:      Move to dest
35:      Wait until 𝒯 ends.
36:      Return to 2
```

Fig. 18 The **SWEEP** cleaning protocol. The term *rightmost neighbor* is defined in Appendix section "The **SWEEP** Cleaning Protocol". *l-d* and *r-d* are left-down and right-down, respectively

criticalness of the pivot point is not necessary for the algorithm to work. Rather, it just makes life easier for the user, as one can now know where to find the agents after the cleaning has been completed. If we do not start with all agents at the pivot and force p_0 to be critical, the location of the agents upon completion will generally not be known in advance.

Fig. 19 Digital sphere of diameter 7, placed around the agent

Signaling

Since by assumption the agent's sensors can detect the status of all tiles which are contained within a digital sphere of radius four placed around the current location of the agent, each agent can artificially calculate the desired destination of all the agents which are located in one of its 4*Neighbors* tiles (see Fig. 19. Thus, the *signaling* action of each agent can be simulated by the other agents near him, and hence an explicit signaling by the agents is not actually required. However, the signaling action is kept in the description and flowchart of the protocol (in procedure 5) for the sake of simplicity and understandability.

Connectivity Preservation

The connectivity of the region yet to be cleaned, F, is preserved by allowing the agents to clean only *non-critical* points. This guarantees both successful termination and agreement of completion (since having no dirty neighbors implies that $F = \emptyset$). Also, should several agents malfunction and halt, as long as there are still functioning agents, the mission will still be carried out, albeit slower.

Note that a tile which was originally clean, or which was cleaned by the agents, is guaranteed to remain clean. As a tile can be cleaned by the agents only when it is in ∂F, it is easy to see that the simple connectivity of F is maintained throughout the cleaning process (as the creation of "holes" — clean tiles which are surrounded by tiles of F — is impossible).

Note that when several agents are located in the same tile, only the last one who exits cleans it (see Line 33 in **SWEEP**), in order to prevent agents from being 'thrown' out of the dirty region (meaning, cleaning a tile in which another agent is located and thus preventing this agent from being able to continue the execution of the cleaning protocol). An alternative method for ensuring the agents are always capable of executing the cleaning protocol would have been the implementation of a "dirtiness seeking mechanism" (i.e. by applying methods such as suggested in [17]). Such a mechanism would have allowed an agent to clean its current location, even if other agents had also been present there. That solution, however, would have been far less elegant and would have added additional difficulties to the analysis process.

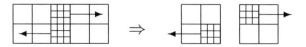

Fig. 20 When the agents do not possess a synchronization mechanism, they may, among others, damage the region's connectivity. In this example, two agents clean their current locations, and move according to the **SWEEP** protocol. Since they are not synchronized, the tiles which they are located in are not treated as critical at the time of cleaning. However, the region's connectivity is not preserved. Should one of the agents had waited for its neighbor to complete executing the protocol's steps before resuming its actions, while deciding whether to clean its current location, it would have treated this tile as critical, and therefore avoid cleaning it. In this case, the connectivity of the region would have been maintained

Agents Synchronization

Note that agents operating in the described environment must have some mean of synchronization, which is mandatory in order to prevent agents from operating in the same time — risking to cut the dirty region into several connected components, as shown in Fig. 20.

To ensure such scenarios will not occur, an order between the operating agents must be implemented. Note — throughout the next paragraphs agents which are signaling a *resting* status (see Appendix section "Clustering Problem" for more details) are being disregarded while calculating the dependencies of the agents' movements. The creation of the following order is implemented in Procedure 19 of **SWEEP**:

Definition 8 For agent i, let $P_i \subseteq \{up, down, left, right, right\text{-}down, left\text{-}down\}$ be a set of directions of tiles, in which there are currently agents, which agent i is delayed by (meaning, agent i will not start moving until the agents in these tiles move). Unless stated otherwise, $P_i = \emptyset$.

For agent i which is located in tile (x, y), if $(x - 1, y)$ is a tile in F, which contains an agent, then $P_i \leftarrow P_i \cup \{left\}$. If $(x, y - 1)$ is a tile in F, which contains an agent, then $P_i \leftarrow P_i \cup \{down\}$, and similarly for $(x + 1, y - 1)$ and *right-down* and for $(x - 1, y - 1)$ and *left-down*.

Definition 9 Let $dest_i \in \{up, down, left, right\}$ be the destination agent i might be interested in moving to, after leaving its current location.

Let each agent be equipped with a two bit flag (that may be implemented for example by two small light-bulbs). This flag states the desired destination of the agent. Alternatively, each tile can be treated as a physical tile, in which the agent can move. Thus, the agent can move towards the top side of the tile, which will be equivalent for using the flag in order to signal that it intends to move *up*.

Let each time step be divided into two phases. In phase 1, every agent "signals" the destination it intends to move towards, either by moving to the appropriate side of the tile, or by using the destination flag.

As we defined an artificial rule which states the superiority of $left$ and $down$ over $right$ and up (and internally, of $down$ over $left$), there are several specific scenarios in which this asymmetry should be reversed in order to ensure a proper operation of the agents. Following is such a "dependencies switching" rule: For agent i, which is located in (x, y), if $dest_i = right$ and tile $(x + 1, y)$ contains an agent j, and $dest_j \neq left$, and there are no other agents which are delayed by agent i (i.e. tile $(x - 1, y)$ does not contain an agent l where $dest_l = right$ and tiles $(x, y + 1)$, $(x + 1, y + 1)$, $(x - 1, y + 1)$ do not contain any agent and tile $(x + 1, y)$ does not contain an agent n where $dest_n = left$), then $P_i \leftarrow P_i \cup \{right\}$ and $P_j \leftarrow P_j \setminus \{left\}$. Also, if $dest_i = up$ and tile $(x, y + 1)$ contains an agent k, and $dest_k \neq down$, and there are no other agents which are delayed by agent i (i.e. tile $(x, y - 1)$ does not contain an agent m where $dest_m = up$ and tiles $(x + 1, y)$, $(x + 1, y + 1)$, $(x - 1, y + 1)$ do not contain any agent and tile $(x, y + 1)$ does not contain an agent q where $dest_q = down$), then $P_i \leftarrow P_i \cup \{up\}$ and $P_k \leftarrow P_k \setminus \{down\}$.

At phase 2 of each time step, the agents start to operate in turns, according to the order implied by P_i. This guarantees that the connectivity of the region is kept, since the simultaneous movement of two neighboring agents is prevented.

Notice that deadlocks are impossible — since the basic rule is that every agent is delayed by the agents in its $left$ and $down$ neighbor tiles. Therefore, at any given time, and for every possible group of agents, there exist an agent with the minimal x and y coordinates (which by definition, is not delayed by any other agent of this group). After this agent moves, all the agents which are delayed by it, can now move, and so on. As to the "dependencies switching rule" — let agent i located in tile (x, y) have the minimal x and y values among the agents who had not moved yet, and let $dest_i = up$, and let tile $(x, y + 1)$ contain an agent j such that $dest_j \neq down$. Then although agent i is located below agent j, it will be delayed by it (i.e. $(up \in P_i)$ and $\neq (down \in P_j)$) as long as agent i is not delaying any other agent (as this is the requirement of the "dependencies switching rule"). In this case, we should show that there can not be a cycle of dependencies, which starts at agent j, ends at agent i, and is closed by the dependency of agent i on agent j. Such a circle can not exist since for it to end in agent i, it means that agent i is delaying another agent k. However, this is impossible since agent i is known not to delay any agent (specifically agent k). Hence, circular dependencies are prevented and no deadlock are possible.

Note that while phase 2 is in process, F_t may change due to cleaning by the agents. As a result, the desired destinations of the agents as well as their dependencies forest, must also be dynamic. This is achieved through the repeated recalculation of these values, for every $waiting$ agent. For example, assume agent i to be located in (x, y), and $dest_i = down$ without loss of generality, and let agent j located in tile $(x, y - 1)$ moves out of this tile and cleans it. Then, $dest_i$ naturally change (as the tile (x, y) does no longer belong to F_t, and thus is not a legitimate destination for agent i), and thus, P_i may also change. In this case, agent i should change its "destination signal" and act according to its new $dest_i$ and P_i. This is implemented in **SWEEP** by calculating the waiting agents' destinations and dependencies lists repeatedly, until either all agents have moved or until the time step has ended (meaning that some agents had to change their status to $resting$, and pause until the next time step — see

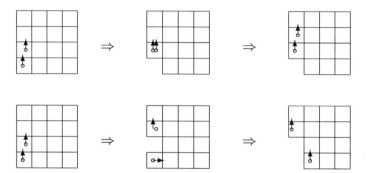

Fig. 21 The upper charts demonstrate a performance bug which may be caused due to local dependencies. The agents are advancing according to the **SWEEP** protocol, but their cleaning performance is decreased. The lower charts demonstrates the cleaning operation after adding the dependencies switching mechanism

Appendix section "Clustering Problem" for more details). Note that every *waiting* agent is guaranteed either to complete its movement in the current time step, or to be forced to wait for the next time step, by switching its status to *resting*.

Notice that the "dependencies switching rule" is not required in order to ensure a proper completion of the mission, but rather to improve the agents' performance, by preventing a bug demonstrated in Fig. 21.

Clustering Problem

Since we are interested in preventing the agents from moving together and functioning as a single agent (due to a resulting decrease in the system's performance), we would like to impose a "resting policy" which will ensure that the agents do not group into clusters which move together. This is done by artificially delaying agents in their current locations, should several agents placed in the same tile wish to move to the same destination. However, we would like the resting time of the agents to be minimal, for performance and analysis reasons.

The following resting policy is implemented by Line 6 of **SWEEP**. Using its sensors, an agent intending to move into tile v is aware of other agents which intend to move into v as well. Thus, when an agent enters a tile which already contains other agents, it knows whether these agents entered this tile at the same time step it did, or whether they had occupied this tile before the current time step had started.

Note that in phase 1 of each time step all the agents are signaling their desired destinations. Thus, an agent can identify which ones of the agents which are located in its current tile intend to move to the same destination as its. Only those agents may cause this agent to delay its actions and rest.

From the agents which intend to move to the same direction as agent i, the agent can distinguish between three groups of agents — the agents which entered this tile before the time step agent i did (group A_2), those which entered the tile in the same time step as i (group A_4), and those who entered it after agent i did (group A_3). If $A_2 \neq \emptyset$, agent i waits, change its status to *resting agent*, signals this status to the other agents in its vicinity, and does not move. As this rule is kept by every other agent, agent i is guaranteed that the agents of group A_3 in their turn will wait for agent i moves before being free to do so as well.

As to the agents of A_4, the problem is solved by induction over the time steps, namely — that at any given time, two agents can not leave the same tile towards the same direction at the same time step. The base of this induction holds for small values of t as the agents are periodically released by the initialization procedure of **SWEEP**, while no two agents are released at the same time step. Later, as we are assured that all the agents of group A_4 arrived to v from different tiles (which are also different than the tile agent i had entered v from), then since all the agents in group A_4 know the previous locations of each other, a consensus over a local ordering of A_4 is established (note that no explicit communication is needed to form this ordering). An example for such an order is the **Priority** function of the **SWEEP** protocol. As a result, the agents are able to exit the tile they are currently located in, in an orderly fashion, according to a well defined order. Thus, the following invariant holds: "*at any given time t, for any two tiles v, u, there can only be a single agent which moves from v to u at time step t*". Thus, the clustering problem is solved.

Notice that there is a single exception to this mechanism, in which the resting commands are overruled. This happens when all non-critical perimeter tiles contain at least two agents. In this scenario, following the resting commands would have created a situation in which no tile can be cleaned, as for every agent which leaves a certain tile, there are still "resting" agents located in the same tile. Therefore, once this scenario is detected by the *check near completion of mission* procedure of **SWEEP**, the agents ignore their resting commands momentarily, in order to be able to clean the non-critical perimeter tiles. The order and internal prioritization of the agents are maintained, for calculating the new *resting* commands in the following time step.

Mission Termination

The termination of the protocol is done in one of two cases — either an agent finds itself in the pivot point, while all of its neighbors are clean (which means that this is the last dirty tile and therefore should be cleaned), or when each dirty tile contains at least one agent (which is a generalization of the previous scenario). The second case is implemented by allowing the agents to signal to their four neighbors whether all of their dirty neighbors contain at least a single agent.

References

1. M. Abramowitz, I.A. Stegun, *Handbook of Mathematical Functions*, Applied Mathematics Series (National Bureau of Standards, 1964), p. 55
2. M. Ahmadi, P. Stone, A multi-robot system for continuous area sweeping tasks, in *IEEE International Conference on Robotics and Automation, 2006 (ICRA 2006)* (2006)
3. S. Alpern, S. Gal, *The Theory of Search Games and Rendezvous* (Kluwer Academic Publishers, Boston, 2003)
4. Y. Altshuler, *Multi Agents Robotics in Dynamic Environments*. Ph.D. thesis, Israeli Institute of Technology, May 2010
5. Y. Altshuler, A.M. Bruckstein, The complexity of grid coverage by swarm robotics, in *ANTS 2010* (LNCS, 2010), pp. 536–543
6. Y. Altshuler, A.M. Bruckstein, Static and expanding grid coverage with ant robots: complexity results. Theoret. Comput. Sci. **412**(35), 4661–4674 (2011)
7. Y. Altshuler, A.M. Bruckstein, I.A. Wagner, Swarm robotics for a dynamic cleaning problem, in *IEEE Swarm Intelligence Symposium* (2005), pp. 209–216
8. Y. Altshuler, V. Yanovski, I.A. Wagner, A.M. Bruckstein, The cooperative hunters - efficient cooperative search for smart targets using uav swarms, in *Second International Conference on Informatics in Control, Automation and Robotics (ICINCO), the First International Workshop on Multi-Agent Robotic Systems (MARS)* (2005), pp. 165–170
9. Y. Altshuler, I.A. Wagner, A.M. Bruckstein, Shape factor's effect on a dynamic cleaners swarm, in *Third International Conference on Informatics in Control, Automation and Robotics (ICINCO), the Second International Workshop on Multi-Agent Robotic Systems (MARS)* (2006), pp. 13–21
10. Y. Altshuler, V. Yanovsky, I. Wagner, A. Bruckstein, Swarm intelligencesearchers, cleaners and hunters. *Swarm Intelligent Systems* (2006), pp. 93–132
11. Y. Altshuler, V. Yanovsky, A.M. Bruckstein, I.A. Wagner, Efficient cooperative search of smart targets using uav swarms. ROBOTICA **26**, 551–557 (2008)
12. Y. Altshuler, I.A. Wagner, A.M. Bruckstein, On swarm optimality in dynamic and symmetric environments. Economics **7**, 11 (2008)
13. Y. Altshuler, I.A. Wagner, A.M. Bruckstein, Collaborative exploration in grid domains, in *Sixth International Conference on Informatics in Control, Automation and Robotics (ICINCO)* (2009)
14. Y. Altshuler, I.A. Wagner, V. Yanovski, A.M. Bruckstein, Multi-agent cooperative cleaning of expanding domains. Int. J. Robot. Res. **30**, 1037–1071 (2010)
15. Y. Altshuler, S. Dolev, Y. Elovici, N. Aharony, Ttled random walks for collaborative monitoring, in NetSciCom, Second International Workshop on Network Science for Communication Networks. San Diego, CA, USA, 3 (2010)
16. Y. Altshuler, A. Pentland, S. Bekhor, Y. Shiftan, A. Bruckstein, Optimal dynamic coverage infrastructure for large-scale fleets of reconnaissance uavs (2016), arXiv:1611.05735
17. R. Baeza-Yates, R. Schott, Parallel searching in the plane. Comput. Geom. **5**, 143–154 (1995)
18. R. Bejar, B. Krishnamachari, C. Gomes, B. Selman, Distributed constraint satisfaction in a wireless sensor tracking system, in *Proceedings of the IJCAI-01 Workshop on Distributed Constraint Reasoning* (2001)
19. M. Bender, S. Michel, P. Triantafillou, G. Weikum, C. Zimmer, Minerva: collaborative p2p search, in *Proceedings of the 31st international conference on Very large data bases* (2005), pp. 1263–1266
20. D.W. Casbeer, D.B. Kingston, R.W. Beard, T.W. McLain, Cooperative forest fire surveillance using a team of small unmanned air vehicles. Int. J. Syst. Sci. **37**(6), 351–360 (2006)
21. M.P. Do-Carmo, *Differential Geometry of Curves and Surfaces* (Prentice-Hall, New-Jersey, 1976)
22. Y. Gabriely, E. Rimon, Competitive on-line coverage of grid environments by a mobile robot. Comput. Geom. **24**, 197–224 (2003)

23. S. Hedberg, Robots cleaning up hazardous waste, in *AI Expert* (1995), pp. 20–24
24. M. Held, On the computational geometry of pocket machining, in *Lecture Notes in Computer Science* (1991)
25. A. Itai, C.H. Papadimitriou, J.L. Szwarefiter, Hamilton paths in grid graphs. SIAM J. Comput. **11**, 676–686 (1982)
26. R. Janakiraman, M. Waldvogel, Q. Zhang, Indra: a peer-to-peer approach to network intrusion detection and prevention, in *Proceedings of IEEE WETICE* (2003)
27. W. Kerr, D. Spears, Robotic simulation of gases for a surveillance task, in *Intelligent Robots and Systems (IROS 2005)* (2005), pp. 2905–2910
28. B.O. Koopman, The theory of search ii, target detection. Oper. Res. **4**(5), 503–531 (1956)
29. D. Latimer, S. Srinivasa, V. Lee-Shue, S. Sonne, H. Choset, A. Hurst, Toward sensor based coverage with robot teams, in *Proceedings of the 2002 IEEE International Conference on Robotics and Automation* (2002)
30. T.W. Min, H.K. Yin, A decentralized approach for cooperative sweeping by multiple mobile robots, in *IEEE/RSJ International Conference on Intelligent Robots and Systems* (1998), pp. 380–385
31. W. Neubauer, Locomotion with articulated legs in pipes or ducts. Robot. Auton. Syst. **11**, 163–169 (1993)
32. K. Passino, M. Polycarpou, D. Jacques, M. Pachter, Y. Liu, Y. Yang, M. Flint, M. Baum, *Cooperative Control for Autonomous Air Vehicles, chapter Cooperative Control and Optimization* (Kluwer Academic, Boston, 2002)
33. M. Polycarpou, Y. Yang, K. Passino, A cooperative search framework for distributed agents, in *IEEE International Symposium on Intelligent Control* (2001)
34. E. Regev, Y. Altshuler, A.M. Bruckstein, The cooperative cleaners problem in stochastic dynamic environments (2012), arXiv:1201.6322
35. I. Rekleitis, V. Lee-Shuey, A. Peng Newz, H. Choset, Limited communication, multi-robot team based coverage, in *IEEE International Conference on Robotics and Automation* (2004)
36. Y. Shoham, M. Tennenholtz, On social laws for artificial agent societies: off line design. AI J. **73**(1–2), 231–252 (1995)
37. L.D. Stone, *Theory of Optimal Search* (Academic Press, New York, 1975)
38. The MathWorks Inc. Matlab — the language of technical computing, ver. 6.5 (2002)
39. I.A. Wagner, Y. Altshuler, V. Yanovski, A.M. Bruckstein, Cooperative cleaners: a study in ant robotics. Int. J. Robot. Res. (IJRR) **27**(1), 127–151 (2008)

The Search Complexity of Collaborative Swarms in Expanding Z^2 Grid Regions

1 Introduction

Motivation. In nature, ants, bees or birds often cooperate to achieve common goals and exhibit amazing feats of swarming behavior and collaborative problem solving. It seems that these animals are "programmed" to interact locally in such a way that the desired global behavior will emerge even if some individuals of the colony die or fail to carry out their task for some other reasons. It is suggested to consider a similar approach to coordinate a group of robots without a central supervisor, by using only local interactions between the robots. When this decentralized approach is used much of the communication overhead (characteristic to centralized systems) is saved, the hardware of the robots can be fairly simple, and better modularity is achieved. A properly designed system should be readily scalable, achieving reliability and robustness through redundancy.

Multi-Agent Robotics and Swarm Robotics. Significant research effort has been invested during the last few years in design and simulation of multi-agent robotics and intelligent swarm systems, e.g. [39, 65, 86].

Swarm based robotic systems can generally be defined as highly decentralized collectives, i.e. groups of extremely simple robotic agents, with limited communication, computation and sensing abilities, designed to be deployed together in order to accomplish various tasks.

Tasks that have been of particular interest to researchers in recent years include synergetic mission planning [6, 42], patrolling [3, 4], fault-tolerant cooperation [58, 88], network security [20–22, 74, 75], swarm control [38, 66], design of mission plans [63, 64], role assignment [35, 36, 43, 91], multi-robot path planning [4, 79, 89], traffic control [5, 73], marketing optimization [23, 24], trends analysis

---

This chapter is based on work previously published in parts in [16, 17, 19].

Y. Altshuler et al., *Swarms and Network Intelligence in Search*,
Studies in Computational Intelligence 729, DOI 10.1007/978-3-319-63604-7_5

[8, 61, 70], formation generation [29, 45, 47], formation keeping [25, 27], exploration and mapping [72, 77, 78], target tracking [48, 71] and distributed search, intruder detection and surveillance [49, 60].

Unfortunately, the mathematical/geometrical theory of such multi agent systems is far from being satisfactory, as pointed out in [28, 30, 41, 69] and many other papers.

Multi Robotics in Dynamic Environments. The vast majority of the works mentioned above discuss challenges involving a multi agent system operating on static domains. Such models, however, are often too limited to capture "real world" problems which, in many cases, involve external element, which may influence their environment, activities and goals. Designing robotic agents that can operate in such environments presents a variety of mathematical challenges.

The main difference between algorithms designed for static environments and algorithms designed to work in dynamic environments is the fact that the agents' knowledge base (either central or decentralized) becomes unreliable, due to the changes that take place in the environment. Task allocation, cellular decomposition, domain learning and other approaches often used by multi agents systems — all become impractical, at least to some extent. Hence, the agents' behavior must ensure that the agents generate a desired effect, regardless the changing environment.

One example is the use of multi agents for distributed search. While many works discuss search after "idle targets", recent works considered dynamic targets, meaning targets which while being searched for by the searching robots, respond by performing various evasive maneuvers intended to prevent their interception. This problem, dating back to World War II operations research (see e.g. [68, 82]), requires the robotic agents to cope with a search area that expands while scanned. The first documented example for search in dynamic domains discussed a planar search problem, considering the scanning of a corridor between parallel borders. This problem was solved in [56] in order to determine optimal strategies for aircraft searching for moving ships in a channel.

A similar problem was presented in [84], where a system consisting of a swarm of UAVs (Unmanned Air Vehicles) was designed to search for one or more "smart targets" (representing for example enemy units, or alternatively a lost friendly unit which should be found and rescued). In this problem the objective of the UAVs is to find the targets in the shortest time possible. While the swarm comprises relatively simple UAVs, lacking prior knowledge of the initial positions of the targets, the targets are assumed to be adversarial and equipped with strong sensors, capable of telling the locations of the UAVs from very long distances. The search strategy suggested in [84] defines *flying patterns* for the UAVs to follow, designed for scanning the (rectangular) area in such a way that the targets cannot re-enter areas which were already scanned by the swarm without being detected. This problem was further discussed in [15], where an improved decentralized search strategy was discussed, demonstrating nearly optimal completion time, compared to the theoretical optimum achievable by any search algorithm.

Collaborative Coverage of Expanding Domains. In this paper we shall examine a problem in which a group of ant-like robotic agents must cover an unknown region in the grid, that possibly expands over time. This problem is also strongly related to the problem of distributed search after mobile and evading target(s) [15, 31, 55] or the problems discussed under the names of "Cops and Robbers" or "Lions and Men" pursuits [44, 46, 50, 51].

We analyze such issues using the results presented in [7, 12, 13, 18, 85], concerning the *Cooperative Cleaners* problem, a problem that assumes a regular grid of connected 'pixels'/'tiles'/'squares'/'rooms', part of which are 'dirty', the 'dirty' pixels forming a connected region of the grid. On this dirty grid region several agents move, each having the ability to 'clean' the place (the 'room', 'tile', 'pixel' or 'square') it is located in. In the dynamic variant of this problem a deterministic evolution of the environment in assumed, simulating a spreading *contamination* (or spreading *fire*). In the spirit of [32] we consider simple robots with only a bounded amount of memory (i.e. *finite-state-machines*).

First, we discuss the collaborative coverage of static grids. We demonstrate that the best completion time known to date ($O(n^2)$, achievable for example using the *LRTA*∗ search algorithm [57]) can be improved to guarantee grid coverage in $O(\frac{1}{k}n^{1.5} + n)$.

Later, we discuss the problem of covering an expanding domain, namely — a region in which "covered" tiles that are adjacent to "uncovered" tiles become "uncovered" every once in a while. Note that the grid is infinite, namely although initial size of the region is n, it can become much greater over time. We show that using any conceivable algorithm, and using as sophisticated and potent robotic agents as possible, the minimal number of robots below which covering such a region is impossible equals $\Omega(\sqrt{n})$. We then show that when the region expands sufficiently slow, specifically — every $O(\frac{c_0}{\gamma_1})$ time steps (where c_0 is the circumference of the region and where γ_1 is a geometric property of the region, which ranges between $O(1)$ and $O(\ln n)$), a group of $\Theta(\sqrt{n})$ robots can successfully cover the region. Furthermore, we demonstrate that in this case a cover time of $O(n^2 \ln n)$ can be guaranteed.

These results are the first analytic results ever concerning the complexity of the number of robots required to cover an expanding grid, as well as for the time such a coverage requires.

2 Related Work

In general, most of the techniques used for the task of a distributed coverage use some sort of cellular decomposition. For example, in [76] the area to be covered is divided between the agents based on their relative locations. In [34] a different decomposition method is being used, which is analytically shown to guarantee a complete coverage of the area. Another interesting work is presented in [2], discussing two methods for cooperative coverage (one probabilistic and the other based on an exact cellular decomposition). All of the works mentioned above, however, rely on the assumption that the cellular decomposition of the area is possible. This in turn, requires the use of memory resources, used for storing the dynamic map generated, the boundaries of the cells, etc'. As the initial size and geometric features of the area are generally not assumed to be known in advance, agents equipped with merely a constant amount of memory will not be able to use such algorithms.

Surprisingly, while some existing works concerning distributed (and decentralized) coverage present analytic proofs for the ability of the system to guarantee the completion of the task (for example, in [2, 26, 34]), most of them lack analytic bounds for the coverage time (although in many cases an extensive amount of empirical results of this nature are made available by extensive simulations).

An interesting work discussing a decentralized coverage of terrains is presented in [90]. This work examines domains with non-uniform traversability. Completion times are given for the proposed algorithm, which is a generalization of the forest search algorithm [92]. In this work, though, the region to be searched is assumed to be known in advance — a crucial assumption for the search algorithm, which relies on a cell-decomposition procedure.

A search for analytic results concerning the completion time of ant-robots covering an area in the grid revealed only a handful of works. The main result in this regard is that of [53, 80], where a swarm of ant-like robots is used for repeatedly covering an unknown area, using a real time search method called *node counting*. By using this method, the robots are shown to be able to efficiently perform such a coverage mission (using integer markers that are placed on the graph's nodes), and analytic bounds for the coverage time are discussed. Based on a more general result for strongly connected undirected graphs shown in [54, 81], the cover time of teams of ant robots (of a given size) that use node counting is shown to be $t_k(n) = O(n^{\sqrt{n}})$, when $t_k(n)$ the cover time of a region of size n using k robots. It should be mentioned though, that in [80] the authors clearly state that it is their belief that the coverage time for robots using node counting in grids is much smaller. This evaluation is also demonstrated experimentally. However, no analytic evidence for this was available thus far.

Another algorithm to be mentioned in this scope is the *LRTA** search algorithm. This algorithm was first introduced in [57] and its multi-robotics variant is shown in [54] to guarantee cover time of undirected connected graphs in polynomial time. Specifically, on grids this algorithm is shown to guarantee coverage in $O(n^2)$ time (again, using integer markers).

Vertex-Ant-Walk, a variant of the node counting algorithm is presented in [87], is shown to achieve a coverage time of $O(n\delta_G)$, where δ_G is the graph's diameter. Specifically, the cover time of regions in the grid is expected to be $O(n^2)$ (however for various "round" regions, a cover time of approximately $O(n^{1.5})$ can be achieved). This work is based on a previous work in which a cover time of $O(n^2\delta_G)$ was demonstrated [83]. Additional analysis of the effect various topological and geometric features of the search space have on the ability of a collaborative swarm to efficiently guarantee successful completion can be found in [9–11].

Another work called *Exploration as Graph Construction*, provides a coverage of degree bounded graphs in $O(n^2)$ time, can be found in [40]. Here a group of limited ant robots explore an unknown graph using special "markers".

Fig. 1 An illustration of a
digital sphere of diameter 7,
placed around a robot

Interestingly, a similar performance can be obtained by using the simplest algorithm for multi robots navigation, namely — random walk. Although in general undirected graphs a group of k random walking robots may require up to $O(n^3)$ time, in
degree bounded undirected graphs such robots would achieve a much faster covering,
and more precisely, $O\left(\frac{|E|^2 log^3 n}{k^2}\right)$ [33]. For regular graphs or degree bounded planar
graphs a coverage time of $O(n^2)$ can be achieved [62], although in such case there
is also a lower bound for the coverage time, which equals $\Omega(n(\log n)^2)$ [52].

We next show that the problem of collaborative coverage in static grid domains
can be completed in $O(\frac{1}{k}n^{1.5} + n)$ time and that collaborative coverage of dynamic
grid domains can be achieved in $O(n^2 \ln n)$ time.

3 The Dynamic Cooperative Cleaners Problem

Following is a short summary of the *Cooperative Cleaners* problem, as appears in
[85] (static variant) and [7, 12] (dynamic variant).

We shall assume that the time is discrete. Let the undirected graph $G(V, E)$ denote
a two dimensional integer grid $\mathbf{Z}^2$, whose vertices (or "*tiles*") have a binary property
called '*contamination*'. Let $cont_t(v)$ state the contamination state of the tile v at time
t, taking either the value "*on*" or "*off*". Let F_t be the contaminated sub-graph of G at
time t, i.e.: $F_t = \{v \in G \mid cont_t(v) = on\}$. We assume that F_0 is a single connected
component. Our algorithm will preserve this property along its evolution.

Let a group of k robots that can move on the grid G (moving from a tile to its
neighbor in one time step) be placed at time t_0 on F_0, at point $p_0 \in F_t$. Each robot is
equipped with a sensor capable of telling the contamination status of all tiles in the
digital sphere of diameter 7, surrounding the robot (namely, in all the tiles that their
Manhattan distance from the robot is equal or smaller than 3. See an illustration in
Fig. 1). A robot is also aware of other robots which are located in these tiles, and all
the robots agree on a common direction. Each tile may contain any number of robots
simultaneously. Each robot is equipped with a memory of size $O(1)$ bits.[1] When a

---

[1]For counting purposes the agents must be equipped with counters that can store the number of
agents in their immediate vicinity. This can of course be implemented using $O(\log k)$ memory.
However, throughout the proof of Lemma 5 in [85] it is shown that the maximal number of agents
that may simultaneously reside in the same tile at any given moment is upper bounded by $O(1)$.
Therefore, counting the agents in the immediate vicinity can be done using counters of $O(1)$ bits.

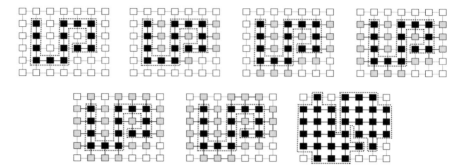

Fig. 2 A demonstration of the barrier expansion process as a result of a contamination spread

robot moves to a tile v, it has the possibility of cleaning this tile (i.e. causing $cont(v)$ to become *off*. The robots do not have any prior knowledge of the shape or size of the sub-graph F_0 except that it is a single and simply connected component.

The contaminated region F_t is assumed to be surrounded at its boundary by a rubber-like elastic barrier, dynamically reshaping itself to fit the evolution of the contaminated region over time. This barrier is intended to guarantee the preservation of the simple connectivity of F_t, crucial for the operation of the robots, due to their limited memory. When a robot cleans a contaminated tile, the barrier retreats, in order to fit the void previously occupied by the cleaned tile. Every d time steps, the contamination spreads. That is, if $t = nd$ for some positive integer n, then:

$$\forall v \in F_t \; \forall u \in 4 - Neighbors(v), \; cont_{t+1}(u) = on$$

Here, the term $4 - Neighbors(v)$ simply means the four tiles adjacent to tile v (namely, the tiles whose Manhattan distance from v equals 1). While the contamination spreads, the elastic barrier stretches while preserving the simple-connectivity of the region, as demonstrated in Fig. 2. For the robots who travel along the tiles of F, the barrier signals the boundary of the contaminated region.

The robots' goal is to clean G by eliminating the contamination entirely.

It is important to note that no central control is allowed, and that the system is fully decentralized (i.e. all robots are identical and no explicit communication between the robots is allowed). An important advantage of this approach, in addition to the simplicity of the robots, is fault-tolerance — even if almost all the robots die and evaporate before completion, the remaining ones will eventually complete the mission, if possible.

A Survey of Previous Results. The cooperative cleaners problem was previously studied in [85] (static version) and [7, 12] (dynamic version). A cleaning algorithm called **SWEEP** was proposed (used by a decentralized group of simple mobile robots, for exploring and cleaning an unknown "contaminated" sub-grid F, expanding every d time steps) and its performance analyzed.

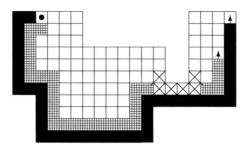

Fig. 3 An example of two agents using the **SWEEP** protocol, at time step 40 (with contamination spreading speed $d > 40$). All the tiles presented were contaminated at time 0. The *black dot* denotes the starting point of the agents. The X's mark the *critical points* which are not cleaned. The *black tiles* are the tiles cleaned by the first agent. The second layer of marked tiles represent the tiles cleaned by the second agent

The **SWEEP** algorithm is based in a constant traversal of the contaminated region, preserving the connectivity of the region while cleaning all *non critical points* — points which when cleaned disconnect the contaminated region. Using this algorithm the agents are guaranteed to stop only upon completing their mission. The algorithm can be implemented using only local knowledge, and local interactions by immediately adjacent agents. At each time step, each agent cleans its current location (assuming it is not a critical point), and moves according to a local movement rule, creating the effect of a clockwise "sweeping" traversal along the boundary of the contaminated region. As a result, the agents "peel" layers from the region, while preserving its connectivity, until the region is cleaned entirely. An illustration of two agents working according to the protocol can be seen in Fig. 3.

In order to formally describe the **SWEEP** algorithm, we should first define several additional terms. Let $\tau(t) = \big(\tau_1(t), \tau_2(t), \ldots, \tau_k(t)\big)$ denote the locations of the k agents at time t. In addition, let $\tilde{\tau}_i(t)$ denote the "previous location" of agent i. Namely, the last tile that agent i had been at, which is different than $\tau_i(t)$. This is formally defined as:

$$\tilde{\tau}_i(t) \triangleq \tau_i(x) \; s.t. x = \max\{j \in \mathbb{N} \mid j < t \; and \; \tau_i(j) \neq \tau_i(t)\}$$

The term ∂F denotes the boundary of F, defined via:

$$\partial F = \{v \mid v \in F \wedge 8 - Neighbors(v) \cap (G \setminus F) \neq \emptyset\}$$

The term '*rightmost*' can now be defined as follows:

- If $t = 0$ then select the tile as instructed in Fig. 4.
- If $\tilde{\tau}_i(t) \in \partial F_t$ then starting from $\tilde{\tau}_i(t)$ (namely, the previous boundary tile that the agent had been in) scan the *four neighbors* of $\tau_i(t)$ in a clockwise order until a boundary tile (excluding $\tilde{\tau}_i(t)$) is found.
- If not $\tilde{\tau}_i(t) \in \partial F_t$ then starting from $\tilde{\tau}_i(t)$ scan the *four neighbors* of $\tau_i(t)$ in a clockwise order until the second boundary tile is found.

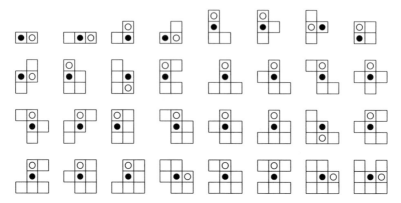

Fig. 4 When $t = 0$ the first movement of an agent located in (x, y) should be decided according to initial contamination status of the neighbors of (x, y), as appears in these charts — the agent's initial location is marked with a filled circle while the destination is marked with an empty one. All configurations which do not appear in these charts can be obtained by using rotations. This definition is needed in order to initialize the traversal behavior of the agents in the correct direction

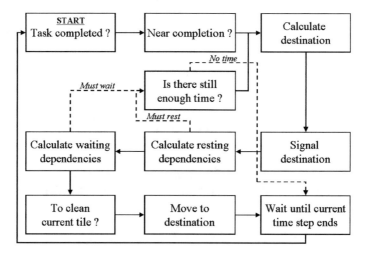

Fig. 5 A schematic flow chart of the **SWEEP** protocol. The smooth lines represent the basic flow of the protocol while the *dashed lines* represent cases in which the flow is interrupted. Such interruptions occur when an agent calculates that it must not move until other agents do so (either as a result of *waiting* or *resting* dependencies — see lines 6 and 14 of **SWEEP** for more details)

A schematic flowchart of the protocol, describing its major components and procedures is presented in Fig. 5. The complete pseudo-code of the protocol and its sub-routines appears in Figs. 6 and 7. Upon initialization of the system, the *System Initialization* procedure is called (defined in Fig. 6). This procedure sets various initial values of the agents, and calls the protocol's main procedure — *SWEEP* (defined in Fig. 7). This procedure in turn, uses various sub-routines and functions, all defined

```
 1: System Initialization
 2:      Arbitrarily choose a pivot point p_0 in ∂F_0, and mark it as critical point
 3:      Place all the agents on p_0
 4:      For (i = 1; i ≤ k; i + +) do
 5:          Call Agent Reset for agent i
 6:          Call SWEEP for agent i
 7:          Wait two time steps
 8:      End for
 9: End procedure

10: Agent Reset
11:      resting ← false
12:      dest ← null /* destination */
13:      near completion ← false
14:      saturated perimeter ← false
15:      waiting ← ∅
16: End procedure

17: Priority
18:      /* Assuming the agent moved from (x_0, y_0) to (x_1, y_1) */
19:      priority ← 2(x_1 − x_0) + (y_1 − y_0)
20: End procedure

21: Check Completion of Mission
22:      If ((x, y) = p_0) and (x, y) has no contaminated neighbors then
23:          If (x, y) is contaminated then
24:              Clean (x, y)
25:          STOP
26: End procedure

27: Check "Near Completion" of Mission
28:      /* Cases where every tile in F_t contains at least a single agent */
29:      near completion ← false
30:      If each of the contaminated neighbors of (x,y) contains at least one agent then
31:          near completion ← true
32:      If each of the contaminated neighbors of (x,y) satisfies near completion then
33:          Clean (x, y) and STOP
34:      /* Cases where every non-critical tile in ∂F_t contains at least 2 agents */
35:      saturated perimeter ← false
36:      If ((x, y) ∈ ∂F_t) and both (x, y) and all of its non-critical neighbors
             in ∂F_t contain at least two agents then
37:          saturated perimeter ← true
38:      If ((x, y) ∈ ∂F_t) and both (x, y) and all of its neighbors in ∂F_t has
             saturated perimeter = true then
39:          Ignore resting commands for this time step
40: End procedure
```

Fig. 6 The first part of the **SWEEP** cleaning protocol

in Fig. 6. The *SWEEP* procedure is comprised of a loop which is executed continuously, until detecting one of two possible break conditions. The first, implemented in the *Check Completion of Mission* procedure, is in charge of detecting cases where all the contaminated tiles have been cleaned. The second condition, implemented in the *Check Near Completion of Mission* procedure, is in charge of detecting scenarios in which every contaminated tile contains at least a single agent. In this case, the next operation would be a simultaneous cleaning of the entire contaminated tiles. Until these conditions are met, each agent goes through the following sequence of commands. First each agent calculates its desired destination at the current turn. Then,

1: **SWEEP Protocol** /* Controls agent i after **Agent Reset** */
2:     **Check Completion of Mission**
3:     **Check "Near Completion" of Mission**
4:     $dest \leftarrow$ *rightmost neighbor* of (x,y) /* Calculate destination */
5:     *destination signal bits* $\leftarrow dest$ /* Signaling the desired destination */
6:     /* Calculate resting dependencies (solves agents' clustering problem) */
7:     From all agents in (x,y) except agent i we define be the following groups:
8:         K_1 : Agents signaling towards $dest$ which entered (x,y) before agent i
9:         K_2 : Agents signaling towards $dest$ which entered (x,y) with agent i,
              and with higher *priority* than this of agent i
10:    $resting \leftarrow false$
11:    **If** $(K_1 \neq \emptyset)$ or $(K_2 \neq \emptyset)$ **then**
12:        $resting \leftarrow true$
13:        **If** (current time-step $\mathcal{T}$ did not end yet) then jump to 4 **Else** jump to 30
14:    $waiting \leftarrow \emptyset$ /* Waiting dependencies (agents synchronization) */
15:    Let *active agent* denote a *non-resting* agent which didn't move in $\mathcal{T}$ yet
16:    **If** (x-1,y) $\in F_t$ contains an active agent then $waiting \leftarrow waiting \cup \{left\}$
17:    **If** (x,y-1) $\in F_t$ contains an active agent then $waiting \leftarrow waiting \cup \{down\}$
18:    **If** (x-1,y-1) $\in F_t$ contains an active agent then $waiting \leftarrow waiting \cup \{l\text{-}d\}$
19:    **If** (x+1,y-1) $\in F_t$ contains an active agent then $waiting \leftarrow waiting \cup \{r\text{-}d\}$
20:    **If** $dest$ = right and (x+1,y) contains an active agent j, and $dest_j \neq left$, and
       there are no other agents delayed by agent i (i.e. (x-1,y) does not contain
       active agent l with $dest_l$ =right and no active agents in (x,y+1),(x+1,y+1),
       (x-1,y+1), and (x+1,y) does not contain active agent n with $dest_n = left$),
       then $\left(waiting \leftarrow waiting \cup \{right\}\right)$ and $\left(waiting_j \leftarrow waiting_j \setminus \{left\}\right)$
21:    **If** $dest$ = up and (x,y+1) contains an active agent j, and $dest_j \neq down$, and
       there are no other agents delayed by agent i (i.e. (x,y-1) does not contain
       active agent l with $dest_l$ =up and no active agents in (x+1,y),(x+1,y+1),
       (x-1,y+1), and (x,y+1) does not contain active agent n with $dest_n = down$),
       then $\left(waiting \leftarrow waiting \cup \{up\}\right)$ and $\left(waiting_j \leftarrow waiting_j \setminus \{down\}\right)$
22:    **If** $(waiting \neq \emptyset)$ **then**
23:        **If** ($\mathcal{T}$ has not ended yet) then jump to 4 **Else** jump to 30
24:    /* Decide whether or not (x,y) should be cleaned */
25:    **If** $\neg ((x,y) \in \partial F_t)$ or $((x,y) \equiv p_0)$ or (x,y) has 2 contaminated tiles in its
       $4Neighbors$ which are not connected via a path of contaminated tiles from its
       $8Neighbors$ **then**
26:        (x,y) is an *internal point* or a *critical point* and should not be cleaned
27:    **Else**
28:        Clean (x,y) if and only if it does not still contain other agents
29:    Move to $dest$
30:    Wait until $\mathcal{T}$ ends.
31:    Return to 2

Fig. 7 The **SWEEP** cleaning protocol

each agent calculated whether it should give a priority to another agent located at the same tile, and wishes to move to the same destination. When two or more agents are located at the same tile, and wish to move towards the same direction, the agent who had entered the tile first gets to leave the tile, while the other agents wait. In case several agents had entered the tile at the same time, the priority is determined using the *Priority* function. Before actually moving, each agent who had obtained a permission to move, must now locally synchronize its movement with its neighbors, in order to avoid simultaneous movements which may damage the connectivity of the region. This is done using the *waiting dependencies* mechanism, which is implemented by each agent via an internal positioning of itself in a local ordering of his neighboring agents. When an agent is not delayed by any other agent, it executes its desired movement. It is important to notice that at any given time, *waiting* or *resting* agents may become active again, if the conditions which made them become inactive in the first place, had changed.

Following are several results that we will later use. While using these results, we note that completely *cleaning* a region is at least as strong as *covering* it, as the number of "uncovered" tiles at any given time can be modeled by the number of "contaminated" tiles, since the number of uncovered tiles in the original region to be explored is clearly upper bounded at all times by the number of remaining contaminated tiles that belong to this region.

Results 1 (Cleaning a Non-Expanding Contamination) *The time it takes for a group of K robots using the **SWEEP** algorithm to clean a region F of the grid is at most:*

$$t_{static} \triangleq \frac{8(|\partial F_0| - 1) \cdot (W(F_0) + k)}{k} + 2k$$

Here $W(F)$ denotes the depth of the region F (the shortest path from some internal point in F to its boundary, for the internal point whose shortest path is the longest) and as defined above, ∂F denotes the boundary of F, defined via:

$$\partial F = \{v \mid v \in F \land 8 - Neighbors(v) \cap (G \setminus F) \neq \emptyset\}$$

The term $8 - Neighbors(v)$ is used to denote the eight tiles that tile v is immediately surrounded by.

Results 2 (Universal Lower Bound on Contaminated Area) *Using any cleaning algorithm, the area at time t of a contaminated region that expands every d time steps can be recursively lower bounded, as follows:*

$$S_{t+d} \geq S_t - d \cdot k + \left\lfloor 2\sqrt{2 \cdot (S_t - d \cdot k) - 1} \right\rfloor$$

Here S_t denotes the area of the contaminated region at time t (such that $S_0 = n$).

Results 3 (Upper Bound on Cleaning Time for SWEEP on Expanding Domains) *For a group of k robot using the **SWEEP** algorithm to clean a region F on the grid, that expands every d time steps, the time it takes the robots to clean F is at most d multiplied by the minimal positive value of the following two numbers:*

$$\frac{(A_4 - A_1 A_3) \pm \sqrt{(A_1 A_3 - A_4)^2 - 4A_3(A_2 - A_1 - A_1 A_4)}}{2A_3}$$

where:

$$A_1 = \frac{c_0 + 2 - \gamma_2}{4}, \quad A_2 = \frac{c_0 + 2 + \gamma_2}{4}, \quad A_3 = \frac{8 \cdot \gamma_2}{d \cdot k},$$

$$A_4 = \gamma_1 - \frac{\gamma_2 \cdot \gamma}{d} \quad , \quad \gamma_1 = \psi(1 + A_2) - \psi(1 + A_1)$$

$$\gamma_2 = \sqrt{(c_0 + 2)^2 - 8S_0 + 8}$$

$$\gamma = \frac{8(k + W(F_0))}{k} - \frac{d - 2k}{|\partial F_0| - 1}$$

Here c_0 is the circumference of the initial region F_0, and where $\psi(x)$ is the *Digamma* function (studied in [1]) — the logarithmic derivative of the *Gamma function*, defined as:

$$\psi(x) = \frac{d}{dx} \ln \Gamma(x) = \frac{\Gamma'(x)}{\Gamma(x)}$$

or as:

$$\psi(x) = \int_0^\infty \left(\frac{e^{-t}}{t} - \frac{e^{-xt}}{1 - e^{-t}} \right) dt$$

Note that although $c_0 = O(|\partial F|)$ the actual length of the perimeter of the region can be greater than its cardinality, as several tiles may be traversed more than once. In fact, in [85] it is shown that $c_0 \leq 2 \cdot |\partial F| - 2$.

4 Grid Coverage — Analysis

We note again that when discussing the coverage of regions on the grid, either static or expanding, it is enough to show that the region can be cleaned by the team of robots, as clearly the cleaned sites are always a subset of the visited ones. We first present the cover time of a group of robots operating in non-expanding domains, using the **SWEEP** algorithm.

Theorem 1 *Given a connected region of $S_0 = n$ grid tiles and perimeter c_0, that should be covered by a team of k ant-like robots, the robots can cover it using $O\left(\frac{1}{k} S_0^{1.5} + S_0\right)$ time.*

Proof Since $|\partial F_0| = \Theta(c_0)$, and $W(F_0) = O(\sqrt{S_0})$, recalling Result 1 we can see that:

$$t_k(n) = t_{static}(k) = O\left(\frac{1}{k}\sqrt{S_0} \cdot c_0 + c_0 + k\right)$$

As $c_0 = O(S_0)$ and as for practical reasons we assume that $k < n$ this would equal in the worse case to:

$$t_k(n) = t_{static}(k) = O\left(\frac{1}{k} S_0^{1.5} + S_0\right)$$

We now examine the problem of covering expanding domains. The lower bound for the number of robots required for completing is as follows.

Theorem 2 *Given a region of size $S_0 \geq 3$ tiles, expanding every d time steps, then a team of less than $\frac{\sqrt{S_0}}{d}$ robots cannot clean the region, regardless of the algorithm used.*

Proof Recalling Result 2 we can see that:

$$S_{t+d} - S_t \geq \left\lfloor 2\sqrt{2 \cdot (S_t - d \cdot k) - 1} \right\rfloor - d \cdot k$$

By assigning $k = \frac{\sqrt{S_0}}{d}$ we can see that:

$$\Delta S_t = S_{t+d} - S_t \geq \left\lfloor 2\sqrt{2 \cdot (S_t - \sqrt{S_0}) - 1} \right\rfloor - \sqrt{S_0}$$

For any $S_0 \geq 3$, we see that $\Delta S_0 > 0$. In addition, for every $S_0 \geq 3$ we can see that $\frac{dS_t}{dt} > 0$ for every $t \geq 0$. Therefore, for every $S_0 \geq 3$ the size of the region will be forever growing.

Corollary 1 *Given a region of size S_0 tiles, expanding every d time steps, where $d = O(1)$ w.r.t S_0, then a team of less than $\Omega(\sqrt{S_0})$ robots cannot clean the region, regardless of the algorithm used.*

Theorem 3 *Given a region F of size S_0 tiles, expanding every d time steps, where $R(F)$ is the perimeter of the bounding rectangle of the region F, then a team of k robots that at $t = 0$ are located at the same tile cannot clean the region, regardless of the algorithm used, as long as $d^2 k < \Omega(R(F))$.*

Proof For every $v \in F$ let $l(v)$ denote the maximal distance between v and any of the tiles of F, namely:

$$l(v) = \max\{d(v, u) | u \in F\}$$

Let $C(F) = l(v_c)$ such that $v_c \in F$ is the tile with minimal value of $l(v)$.

Let v_s denote the tile the agents are located in at $t = 0$. Let $v_d \in F$ denote some contaminated tile such that $d(v_s, v_d) = l(v_s)$. Regardless of the algorithm used by the agents, until some agent reaches v_d there will pass at least $l(v_s)$ time steps. Let us assume w.l.o.g that v_d is located to the right (or of the same horizontal coordinate) and to the top (or of the same vertical coordinate) of v_s. Then by the time some agent is able to reach v_d there exists an upper-right quarter of a digital sphere of radius $\left\lfloor \frac{l(v_s)}{d} \right\rfloor + 1$, whose center is v_d.

The number of tiles in such a quarter of digital sphere equals:

$$\frac{1}{2} \left\lfloor \frac{l(v_s)}{d} \right\rfloor^2 + \frac{3}{2} \left\lfloor \frac{l(v_s)}{d} \right\rfloor + 1 = \Theta \left(\frac{l(v_s)^2}{d^2} \right)$$

It is obvious that the region cannot be cleaned until v_d is cleaned. Let t_d denote the time at which the first agent reaches v_d. It is easy to see that $t_d \geq l(v_s)$. Therefore, regardless of activities of the agents until time step t_d, there are now k agents that has to clean a region of at least $\Theta \left(\frac{l(v_s)^2}{d^2} \right)$ tiles, spreading every d time steps. Using Theorem 2 we know that k agents cannot clean an expanding region of $k = \frac{\sqrt{S_0}}{d}$ tiles. Namely, at time t_d k agents could not clean the contaminated tiles if:

$$d^2 k < \Omega \left(l(v_s) \right)$$

As $l(v_s) \geq C(F)$ we know that k agents could not clean an expanding contaminated region where: $d^2 k < \Omega \left(C(F) \right)$. It is easy to see that for every region F, if $R(F)$ is the length of the perimeter of the bounding rectangle of F then $C(F) = \Theta(R(F))$.

Lemma 1 *For every connected region of size $S_0 \geq 3$ and perimeter of length c_0:*

$$\frac{1}{2} c_0 < \gamma_2 < c_0$$

Proof Let us assume by contradiction that $(c_0 + 2)^2 \leq (8 S_0 + 8)$. This means $c_0 \leq \sqrt{8 S_0 + 8} - 2$. However, the minimal circumference of a region of size S_0 is achieved when the region is arranged in the form of an 8-connected digital sphere, in which case $c_0 \geq 4\sqrt{S_0} - 4$, contradicting the assumption that $c_0 \leq \sqrt{8 S_0 + 8} - 2$ for every $S_0 > 5$. Therefore, $\gamma_2 \in \mathbb{R}$.

Let us assume by contradiction that $\gamma_2 < \frac{1}{2} c_0$. Therefore:

$$(c_0 + 2)^2 - 8 S_0 + 8 < \frac{1}{4} c_0^2$$

which implies:

$$c_0 < -\frac{16}{6} + \sqrt{10\frac{2}{3} S_0 - 8\frac{8}{9}} < 3.266\sqrt{S_0} - 2$$

However, we know that $c_0 \geq 4\sqrt{S_0} - 4$, which contradicts the assumption that $\gamma_2 < \frac{1}{2} c_0$ for every $S_0 \geq 3$.

Let us assume by contradiction that $\gamma_2 > c_0$. Therefore:

$$(c_0 + 2)^2 - 8 S_0 + 8 > c_0^2$$

which implies:

$$c_0 > 4S_0 - 6$$

However, we know that $c_0 \leq 2S_0 - 2$ (as c_0 is maximized when the tiles are arranged in the form of a straight line), contradicting the assumption that $\gamma_2 > c_0$ for every $S_0 \geq 3$.

Lemma 2 *For every connected region of size $S_0 \geq 3$ and perimeter of length c_0:*

$$\Omega(1) < \gamma_1 < O(\ln n)$$

Proof Let us observe γ_1:

$$\gamma_1 \triangleq \psi\left(1 + \frac{c_0 + 2 + \gamma_2}{4}\right) - \psi\left(1 + \frac{c_0 + 2 - \gamma_2}{4}\right)$$

From Lemma 1 we can see that $1 < \left(1 + \frac{c_0 + 2 - \gamma_2}{4}\right) < \frac{1}{4}c_0$. Note that $\psi(1) = -\hat{\gamma}$ where $\hat{\gamma}$ is the *Euler?-Mascheroni* constant, defined as:

$$\hat{\gamma} = \lim_{n \to \infty}\left[\left(\sum_{k=1}^{n}\frac{1}{k}\right) - \log(n)\right] = \int_{1}^{\infty}\left(\frac{1}{\lfloor x \rfloor} - \frac{1}{x}\right) dx$$

which equals approximately 0.57721. In addition, $\psi(x)$ is monotonically increasing for every $x > 0$. As we also know that $\psi(x)$ is upper bounded by $O(\ln x)$ for large values of x, we see that:

$$-0.58 < \psi\left(1 + \frac{c_0 + 2 - \gamma_2}{4}\right) < O(\ln n) \tag{1}$$

From Lemma 1 we also see that $1 < \left(1 + \frac{c_0 + 2 + \gamma_2}{4}\right) < \frac{1.5}{4}c_0$ meaning that:

$$\psi\left(1 + \frac{c_0 + 2 + \gamma_2}{4}\right) = \Theta(\ln n) \tag{2}$$

Combining Eqs. 1 and 2 we see that:

$$\Omega(1) < \gamma_1 < O(\ln n) \tag{3}$$

Theorem 4 *Result 3 returns a positive real number for the covering time of a region of S_0 tiles that expands every d time steps, when the number of robots is $\Theta(\sqrt{S_0})$ and $d = \Omega(\frac{c_0}{\gamma_1})$, and where γ_1 shifts from $O(1)$ to $O(\ln S_0)$ as c_0 grows from $O(\sqrt{S_0})$ to $O(S_0)$, defined as:*

$$\gamma_1 = \psi\left(1 + \frac{c_0 + 2 + \gamma_2}{4}\right) - \psi\left(1 + \frac{c_0 + 2 - \gamma_2}{4}\right)$$

$$\gamma_2 = \sqrt{(c_0 + 2)^2 - 8S_0 + 8}$$

Proof Following are the requirements that must hold in order for Result 3 to yield a real number:

- $d \cdot k \neq 0$
- $|\partial F| > 1$
- $A_3 \neq 0$
- $(c_0 + 2)^2 > 8S_0 - 8$
- $(A_1 A_3 - A_4)^2 \geq 4A_3(A_2 - A_1 - A_1 A_4)$

The first and second requirements hold for every non trivial scenario. The third requirement is implied by the fourth. The fourth assumption is a direct result of Lemma 1.

As for the last requirement, we ask that:

$$A_1^2 A_3^2 + A_4^2 \geq 4A_2 A_3 - 4A_1 A_3 - 2A_1 A_3 A_4$$

which subsequently means that we must have:

$$\frac{\gamma_2^2}{d^2 k^2}\left(c_0^2 + \gamma_2^2 - c_0\gamma_2\right) + \gamma_1^2 + \frac{\gamma_2^2 \gamma^2}{d} - \frac{\gamma\gamma_1\gamma_2}{d} \geq 4\frac{\gamma_2}{dk}\left(c_0\frac{\gamma_2\gamma}{d} - 2\gamma_1 + 2\frac{\gamma_2\gamma}{d} + \gamma_1\gamma_2 - \frac{\gamma_2^2\gamma}{d}\right)$$

Using Lemmas 1 and 2 we should make sure that:

$$\frac{\gamma_2^2}{dk^2}c_0^2 + \gamma_1^2 d + \gamma_2^2\gamma^2 - \gamma\gamma_1\gamma_2 \geq O\left(\frac{c_0\gamma_2\gamma_1}{k} + \frac{c_0\gamma_2^2\gamma}{dk}\right)$$

Using $W(F) = O(\sqrt{S_0})$ and $\Omega(\sqrt{S_0}) = |\partial F| = O(S_0)$ and dividing by γ_2^2 (which we know to be larger than zero), we can write the above as follows:

$$\frac{c_0^2}{dk^2} + \frac{k^2 + d\ln^2 S_0}{c_0^2} + 1 \geq$$

$$O\left(\frac{\ln S_0}{c_0} + \frac{k\ln S_0}{c_0^2} + \frac{\ln S_0}{k} + \frac{c_0\sqrt{S_0}}{dk^2} + \frac{c_0}{dk} + \frac{1}{d}\right)$$

As $c_0 \geq \sqrt{S_0}$ then $\frac{c_0^2}{dk^2} \geq \frac{c_0\sqrt{S_0}}{dk^2}$. In addition, $1 \geq \frac{1}{d}$ and also $1 \geq \frac{\ln S_0}{c_0}$ and $1 \geq \frac{\ln S_0}{k}$ (as Eq. 6 shows that $k \geq \ln S_0$). In order to have also $1 \geq \frac{c_0}{dk}$ we must have:

$$d \cdot k = \Omega(c_0) \tag{4}$$

In addition, we should also require that the result μ would be positive (as it denotes the coverage time). Namely, that:

$$A_4 + \sqrt{(A_1 A_3 - A_4)^2 - 4A_3(A_2 - A_1 - A_1 A_4)} > A_1 A_3$$

For this to hold we shall merely require that:

$$A_2 - A_1 - A_1 A_4 \leq 0$$

(as A_3 is known to be positive). Assigning the values of A_1, A_2, A_4, this translates to:

$$c_0 + c_0^2 \frac{\gamma}{d} \leq O(c_0 \gamma_1)$$

Dividing by c_0 we can write:

$$1 + c_0 \frac{1 + \frac{\sqrt{S_0}}{k} - \frac{d}{c_0} + \frac{k}{c_0}}{d} \leq O(\gamma_1)$$

Namely:

$$c_0 + \frac{c_0 \sqrt{S_0}}{k} + k \leq d\, O(\gamma_1)$$

As c_0 is the dominant element of the left side of the inequation, we see that:

$$d = \Omega \left(\frac{c_0}{\gamma_1} \right) \tag{5}$$

Assigning this lower bound for d we can now see that:

$$\Omega(\sqrt{S_0}) \leq k \leq O(c_0) \tag{6}$$

Therefore, we shall select the value of k such that:

$$k = \Theta(\sqrt{S_0})$$

This also satisfies Eq. 4.

Theorem 5 *The time it takes a group of $k = \Theta(\sqrt{S_0})$ robots using the **SWEEP** algorithm to cover a connected region of size S_0 tiles, that expands every $d = \Omega(\frac{c_0}{\gamma_1})$ time steps, is upper bounded as follows:*

$$t_{SUCCESS} = O\left(S_0^2 \ln S_0 \right)$$

where γ_1 shifts from $O(1)$ to $O(\ln S_0)$ as c_0 grows from $O(\sqrt{S_0})$ to $O(S_0)$, defined as:

$$\gamma_1 = \psi\left(1 + \frac{c_0 + 2 + \gamma_2}{4}\right) - \psi\left(1 + \frac{c_0 + 2 - \gamma_2}{4}\right)$$

$$\gamma_2 = \sqrt{(c_0 + 2)^2 - 8S_0 + 8}$$

Proof Recalling Result 1 we know that if:

$$\frac{8(|\partial F_0| - 1) \cdot (W(F_0) + k)}{k} + 2k < d$$

then the robots could clean the region before it expands even once. In this case, the cleaning time would be $O(\frac{1}{k}\sqrt{S_0} \cdot c_0 + c_0)$ as was shown in Theorem 1. Therefore, we shall assume that:

$$\frac{8(|\partial F_0| - 1) \cdot (W(F_0) + k)}{k} + 2k \geq d \tag{7}$$

Observing Result 3 we see that:

$t_{SUCCESS} =$

$$d \cdot O\left(A_1 + \frac{|A_4|}{A_3} + \sqrt{A_1^2 + \frac{|A_1 A_4| + A_1 + A_2}{A_3} + \frac{A_4^2}{A_3^2}}\right) \leq$$

$$d \cdot O\left(A_1 + \frac{|A_4|}{A_3} + \frac{\sqrt{A_3}\sqrt{|A_1 A_4| + A_1 + A_2}}{A_3} + \frac{|A_4|}{A_3}\right) \leq$$

$$d \cdot O\left(A_1 + \frac{|A_4|}{A_3} + \frac{\sqrt{|A_1 A_4|} + \sqrt{A_1} + \sqrt{A_2}}{\sqrt{A_3}}\right) =$$

$$d \cdot O\left(\frac{c_0 + \gamma_2 + dk\frac{\gamma_1}{\gamma_2} + k\gamma +}{\sqrt{k}\frac{\sqrt{c_0 + \gamma_2}\sqrt{d\gamma_1 + \gamma_2 \cdot \gamma}}{\sqrt{\gamma_2}} + \sqrt{\frac{kd}{\gamma_2}}\sqrt{c_0 + \gamma_2}}\right)$$

Using the fact that $\gamma_2 = \Theta(c_0)$ (Lemma 1) we can rewrite this expression as:

$$d \cdot O\left(c_0 + dk\frac{\gamma_1}{c_0} + k\gamma + \sqrt{k}\sqrt{d\gamma_1 + c_0\gamma} + \sqrt{kd}\right) \tag{8}$$

Recalling Eq. 7, and as $W(F_0) = O(\sqrt{S_0})$, we can see that:

$$d = O\left(\frac{\sqrt{S_0} \cdot c_0}{k} + c_0 + k\right)$$

Therefore, $|\gamma|$ can now be written as:

$$|\gamma| = O\left(\frac{\sqrt{S_0}}{k} + \sqrt{S_0} + \frac{k}{\sqrt{S_0}} + 1\right)$$

In addition, remembering that $O(\sqrt{S_0}) \le c_0 \le O(S_0)$ we can rewrite the expression of Eq. 8 as follows:

$$d \cdot O\left(\frac{c_0 + dk\frac{\gamma_1}{c_0} + k\sqrt{S_0} + \frac{k^2}{\sqrt{S_0}} +}{\sqrt{kc_0}\sqrt{\frac{d}{c_0}\gamma_1 + \frac{\sqrt{S_0}}{k} + \sqrt{S_0} + \frac{k}{\sqrt{S_0}}} + \sqrt{kd}}\right) =$$

$$d \cdot O\left(\frac{c_0 + dk\frac{\gamma_1}{c_0} + k\sqrt{S_0} + \frac{k^2}{\sqrt{S_0}} +}{\sqrt{kc_0}\left(\sqrt{\gamma_1} + \sqrt[4]{S_0}\sqrt{\frac{\gamma_1}{k}} + \sqrt[4]{S_0} + \frac{\sqrt{k}}{\sqrt[4]{S_0}}\right) + \sqrt{kd}}\right)$$

Using Lemma 2 we see that:

$$d \cdot O\left(\frac{c_0 + dk\frac{\ln S_0}{c_0} + k\sqrt{S_0} + \frac{k^2}{\sqrt{S_0}} +}{\sqrt{c_0 \ln S_0\sqrt{S_0}} + \sqrt{c_0 k\sqrt{S_0}} + k\sqrt[4]{S_0} + \sqrt{kd}}\right) =$$

$$d \cdot O\left(\frac{c_0 + k\sqrt{S_0}\ln S_0 + \frac{k^2}{\sqrt{S_0}} +}{\sqrt{c_0 \ln S_0\sqrt{S_0}} + \sqrt{c_0 k\sqrt{S_0}}}\right) \tag{9}$$

Assuming that $k > O(\ln S_0)$ and as $c_0 = O(S_0)$ we can now write:

$$O\left(\frac{\frac{S_0^{2.5}}{k} + S_0^2 \ln S_0 + S_0^{1.5}k \ln S_0 + k^2\sqrt{S_0}\ln S_0 +}{\frac{k^3}{\sqrt{S_0}} + \frac{S_0^{2.25}}{\sqrt{k}} + S_0^{1.75}\sqrt{k} + S_0^{0.75}k^{1.5}}\right) \tag{10}$$

Using Eq. 6 we can see that this translates to:

$$O\left(S_0^{1.5}k \ln S_0\right) \tag{11}$$

5 Dynamic Cleaners as Cooperative Hunters

When designing an infrastructure for efficiently scanning an area for evading targets (for example, see works such as [37, 56, 59, 67, 84] or the work that is presented in the next Chapter) the area to be scanned is not necessarily rectangular, or known in advance. Granted that for any non rectangular area of size S one can always use any given search algorithm that is designed for a rectangular area, and apply it on the bounding rectangle of this area. However, it can easily be seen that S_{Rect}, the area of this bounding rectangle, can be significantly larger compared to S. Moreover, for various search areas it can even be seen that $S_{Rect} = O(S^2)$.

Furthermore, even if we are willing to "pay" the price of a significantly larger search area, using this approach still requires us to possess the exact shape of the search area, prior to the launch of the drones. Therefore, as this knowledge is not always available, it is interesting to see whether we can design a more flexible decentralized scanning algorithm.

Interestingly, it can be shown that under some modifications, robots designed for solving the Dynamic Cooperative Cleaners problem can be used to create a decentralized hunting system, for unknown search regions. This reduction is proposed to be done as follows:

First, let us sample the search area from $\mathbb{R}^2$ to the grid $\mathbb{Z}^2$. Let us now denote the sampled area to be scanned by F_0. As the targets' location and quantity are unknown, we shall treat F_0 as a "danger zone", namely — the union of all possible locations of the targets. In addition, we shall assume that the targets to be "as lucky as possible", namely — even if F is reduced to a single tile, then we shall assume that the targets have chosen a path which would have led them to be located at this tile. Therefore, the targets are known to be found, when and only when $F = \emptyset$. As the targets can move, the "danger zone" F expands, proportionally to the velocity of the targets (normalized by the velocity of the UAVs). Namely: $d = \frac{V_{UAV}}{V_{target}}$.

At this point it should be noted that unlike the cooperative cleaners problem, in which the contamination status of the tiles is used by the agents as some form of an implicit communication, when using this approach for a decentralized scanning mission, the agents must maintain some sort of virtual global representation of the "contaminated ares" F at any given time (which means that the agents must be equipped with a stronger means of communication than available in the original model).

Under these assumptions, a swarm containing k drones can be released at the boundary of the search area. Using the results which are presented in Sect. 4, lower and upper bounds of the performance of those drones can then be obtained.

6   Conclusions

In this paper we have discussed the problem of cleaning or covering a connected region on the grid using a collaborate team of simple, finite-state-automata robotic agents. We have shown that when the regions are static, this can be done in $O(\frac{1}{k}\sqrt{n} \cdot c_0 + c_0 + k)$ time which equals $O(\frac{1}{k}n^{1.5} + n)$ time in the worst case, thus improving the previous results for this problem. In addition, we have shown that when the region is expanding in a constant rate (which is "slow enough"), a team of $\Theta(\sqrt{n})$ robots can still be guaranteed to clean or cover it, in $O(n^2 \ln n)$ time.

In addition, we have shown that teams of less than $\Omega(\sqrt{n})$ robots can *never* cover a region that expands every $O(1)$ time steps, regardless of their sensing capabilities, communications and memory resources employed, or the algorithm used. As to regions that expand slower than every $O(1)$ time steps, we have shown the following impossibility results. First, a region of n tiles that expands every d time steps cannot be covered by a group of k agents if $dk \leq O(\sqrt{n})$. Using Theorem 4 we can guarantee a coverage when $dk = \Omega(\frac{n^{1.5}}{\ln n})$, or even for $dk = \Omega(n)$) (when the region's perimeter c_0 equals $O(n)$).

Second, a spreading region cannot be covered when $d^2 k$ is smaller than the order of the perimeter of the bounding rectangle of the region (which equals $O(n)$ in the worse case and $O(\sqrt{n})$ for shapes with small perimeters). Using Theorem 4 we can guarantee a coverage when $d^2 k = \Omega(\frac{n^{2.5}}{\ln^2 n})$, or for $d^2 k = \Omega(n^{1.5})$ (when the region's perimeter c_0 equals $O(n)$).

We believe that these results can be easily applied to various other problems that can be reduced or modeled as a swarm of autonomous agents that are required to operate in an expanding grid domain. For example, this result can show that a team of $\Theta(\sqrt{n})$ cops can always catch a robber (or for that matter — a group of robbers), moving (slowly) in an unbounded grid. Alternatively, robbers can be guaranteed to escape a team of less than $O(\sqrt{n})$ cops, if the area they are located in is unbounded. Similarly, this can rather immediately be applied for swarms of autonomous drones that are required to intercept a group of evading targets in a bounded region [14, 15, 21].

References

1. M. Abramowitz, I.A. Stegun, *Handbook of Mathematical Functions*. Applied Mathematics Series (National Bureau of Standards 1964), p. 55
2. E.U. Acar, Y. Zhang, H. Choset, M. Schervish, A.G. Costa, R. Melamud, D.C. Lean, A. Gravelin, Path planning for robotic demining and development of a test platform, in *International Conference on Field and Service Robotics* (2001), pp. 161–168
3. N. Agmon, S. Kraus, G.A. Kaminka. Multi-robot perimeter patrol in adversarial settings, in *IEEE International Conference on Robotics and Automation (ICRA 2008)* (2008), pp. 2339–2345
4. N. Agmon, V. Sadov, G.A. Kaminka, S. Kraus, The impact of adversarial knowledge on adversarial planning in perimeter patrol, in *AAMAS '08: Proceedings of the 7th International Joint Conference on Autonomous Agents and Multiagent Systems* (International Foundation for Autonomous Agents and Multiagent Systems, Richland, SC, 2008), pp. 55–62
5. A. Agogino, K. Tumer, Regulating air traffic flow with coupled agents, in *AAMAS '08: Proceedings of the 7th International Joint Conference on Autonomous Agents and Multiagent Systems* (International Foundation for Autonomous Agents and Multiagent Systems, Richland, SC, 2008), pp. 535–542
6. R. Alami, S. Fleury, M. Herrb, F. Ingrand, F. Robert, Multi-robot cooperation in the martha project. IEEE Robot. Autom. Mag. **5**(1), 36–47 (1998)
7. Y. Altshuler, A.M. Bruckstein, I.A. Wagner, Swarm robotics for a dynamic cleaning problem, in *IEEE Swarm Intelligence Symposium* (2005), pp. 209–216

8. Y. Altshuler, M. Fire, N. Aharony, Y. Elovici, A Pentland, How many makes a crowd? on the correlation between groups' size and the accuracy of modeling, in *International Conference on Social Computing, Behavioral-Cultural Modeling and Prediction* (Springer, 2012), pp. 43–52

9. Y. Altshuler, I.A. Wagner, A.M. Bruckstein, Shape factor's effect on a dynamic cleaners swarm, in *Third International Conference on Informatics in Control, Automation and Robotics (ICINCO), the Second International Workshop on Multi-Agent Robotic Systems (MARS)* (2006), pp. 13–21

10. Y. Altshuler, I.A. Wagner, A.M. Bruckstein, On swarm optimality in dynamic and symmetric environments. Economics **7**, 11 (2008)

11. Y. Altshuler, I.A. Wagner, A.M. Bruckstein, Collaborative exploration in grid domains, in *Sixth International Conference on Informatics in Control, Automation and Robotics (ICINCO)* (2009)

12. Y. Altshuler, I.A. Wagner, A.M. Bruckstein, Swarm ant robotics for a dynamic cleaning problem — upper bounds, in *The 4th International conference on Autonomous Robots and Agents (ICARA-2009)* (2009), pp. 227–232

13. Y. Altshuler, I.A. Wagner, V. Yanovski, A.M. Bruckstein, Multi-agent cooperative cleaning of expanding domains. Int. J. Robot. Res. **30**, 1037–1071 (2010)

14. Y. Altshuler, V. Yanovski, I.A. Wagner, A.M. Bruckstein, The cooperative hunters - efficient cooperative search for smart targets using uav swarms, in *Second International Conference on Informatics in Control, Automation and Robotics (ICINCO), the First International Workshop on Multi-Agent Robotic Systems (MARS)* (2005), pp. 165–170

15. Y. Altshuler, V. Yanovsky, A.M. Bruckstein, I.A. Wagner, Efficient cooperative search of smart targets using uav swarms. Robotica **26**, 551–557 (2008)

16. Y. Altshuler, V. Yanovsky, I. Wagner, A. Bruckstein. Swarm intelligencesearchers, cleaners and hunters, in *Swarm Intelligent Systems* (2006), pp. 93–132

17. Y. Altshuler, *Multi Agents Robotics in Dynamic Environments*. Ph.D. thesis, Israeli Institute of Technology (2010)

18. Y. Altshuler, A.M. Bruckstein, The complexity of grid coverage by swarm robotics, in *ANTS 2010* (LNCS, 2010), pp. 536–543

19. Y. Altshuler, A.M. Bruckstein, Static and expanding grid coverage with ant robots: complexity results. Theor. Comput. Sci. **412**(35), 4661–4674 (2011)

20. Y. Altshuler, S. Dolev, Y. Elovici, N. Aharony, Ttled random walks for collaborative monitoring, in *NetSciCom, Second International Workshop on Network Science for Communication Networks*, vol. 3 (San Diego, CA, USA, 2010)

21. Y. Altshuler, A. Pentland, S. Bekhor, Y. Shiftan, A. Bruckstein, Optimal dynamic coverage infrastructure for large-scale fleets of reconnaissance uavs (2016), arXiv:1611.05735

22. Y. Altshuler, R. Puzis, Y. Elovici, S. Bekhor, A.S. Pentland, On the rationality and optimality of transportation networks defense: a network centrality approach, in *Securing Transportation Systems*, pp. 35–63

23. Y. Altshuler, E. Shmueli, G. Zyskind, O. Lederman, N. Oliver, A. Pentland, Campaign optimization through behavioral modeling and mobile network analysis. IEEE Trans. Comput. Soc. Syst. **1**(2), 121–134 (2014)

24. Y. Altshuler, E. Shmueli, G. Zyskind, O. Lederman, N. Oliver, A.S. Pentland, Campaign optimization through mobility network analysis, in *Geo-Intelligence and Visualization Through Big Data Trends* (2015), pp. 33–74

25. T. Balch, R. Arkin, Behavior-based formation control for multi-robot teams. IEEE Trans. Robot. Autom. **14**(6), 926–939 (1998)

26. M.A. Batalin, G.S. Sukhatme, Spreading out: a local approach to multi-robot coverage, in *6th International Symposium on Distributed Autonomous Robotics Systems* (2002)

27. K. Bendjilali, F. Belkhouche, B. Belkhouche, Robot formation modelling and control based on the relative kinematics equations. Int. J. Robot. Autom. **24**(1), 79–88 (2009)

28. G. Beni, J. Wang, Theoretical problems for the realization of distributed robotic systems, in *IEEE Internal Conference on Robotics and Automation* (1991), pp. 1914–1920
29. R.M. Bhatt, C.P. Tang, V.N. Krovi, Formation optimization for a fleet of wheeled mobile robots a geometric approach. Robot. Auton.Syst. **57**(1), 102–120 (2009)
30. E. Bonabeau, M. Dorigo, G. Theraulaz, *Swarm Intelligence: From Natural to Artificial Systems* (Oxford University Press, US, 1999)
31. R. Borie, C. Tovey, S. Koenig, Algorithms and complexity results for pursuit-evasion problems, in *Proceedings of the International Joint Conference on Artificial Intelligence (IJCAI)* (2009), pp. 59–66
32. V. Braitenberg, *Vehicles* (MIT Press, Cambridge, 1984)
33. A.Z. Broder, A.R. Karlin, P. Raghavan, E. Upfal, Trading space for time in undirected $s − t$ connectivity, in *ACM Symposium on Theory of Computing (STOC)* (1989), pp. 543–549
34. Z. Butler, A. Rizzi, R. Hollis, Distributed coverage of rectilinear environments, in *Proceedings of the Workshop on the Algorithmic Foundations of Robotics* (2001)
35. C. Candea, H. Hu, L. Iocchi, D. Nardi, M. Piaggio, Coordinating in multi-agent robocup teams. Robot. Auton. Syst. **36**(2–3), 67–86 (2001)
36. G. Chalkiadakis, C. Boutilier, Sequential decision making in repeated coalition formation under uncertainty, in *Proceedings of the 7th international joint conference on Autonomous agents and multiagent systems* (2008), pp. 347–354
37. H. Chung, E. Polak, J.O. Royset, S. Sastry, On the optimal detection of an underwater intruder in a channel using unmanned underwater vehicles. Naval Res. Logist. (NRL), **58**(8), 804–820 (2011)
38. R. Connaughton, P. Schermerhorn, M. Scheutz. Physical parameter optimization in swarms of ultra-low complexity agents, in *AAMAS '08: Proceedings of the 7th International Joint Conference on Autonomous Agents and Multiagent Systems* (International Foundation for Autonomous Agents and Multiagent Systems, Richland, SC, 2008), pp. 1631–1634
39. S.A. DeLoach, M. Kumar, in *Intelligence Integration in Distributed Knowledge Management*, Multi-agent systems engineering: an overview and case study (Idea Group Inc (IGI), 2008), pp. 207–224
40. G. Dudek, M. Jenkin, E. Milios, D. Wilkes, Robotic exploration as graph construction. IEEE Trans. Robot. Autom. **7**, 859–865 (1991)
41. A. Efraim, D. Peleg, Distributed algorithms for partitioning a swarm of autonomous mobile robots. Struct. Inf. Commun. Complex. Lect. Notes Comput. Sci. **4474**, 180–194 (2007)
42. U. Visser et al., *RoboCup 2007: Robot Soccer World Cup XI*, vol. 5001, Lecture Notes in Computer Science (Springer, Berlin, 2008)
43. A. Felner, Y. Shoshani, Y. Altshuler, A.M. Bruckstein, Multi-agent physical a* with large pheromones. J. Auton. Agents Multi-Agent Syst. **12**(1), 3–34 (2006)
44. J.O. Flynn, Lion and man: the general case. SIAM J. Control **12**, 581–597 (1974)
45. J. Fredslund, M.J. Mataric, Robot formations using only local sensing and control, in *Proceedings of the International Symposium on Computational Intelligence in Robotics and Automation (IEEE CIRA 2001)* (2001), pp. 308–313
46. A.S. Goldstein, E.M. Reingold, The complexity of pursuit on a graph. Theor. Comput. Sci. **143**, 93–112 (1995)
47. N. Gordon, I.A. Wagner, A.M. Bruckstein. Discrete bee dance algorithms for pattern formation on a grid, in *IEEE International Conference on Intelligent Agent Technology (IAT03)* (2003), pp. 545–549
48. I. Harmatia, K. Skrzypczykb, Robot team coordination for target tracking using fuzzy logic controller in game theoretic framework. Robot. Auton. Syst. **57**(1), 75–86 (2009)
49. G. Hollinger, S. Singh, J. Djugash, A. Kehagias, Efficient multi-robot search for a moving target. Int. J. Robot. Res. **28**(2), 201–219 (2009)
50. R.P. Isaacs, *Differential Games: A Mathematical Theory with Applications to Warfare and Pursuit, Control and Optimization* (Wiley, New York, 1965). Reprinted by Dover Publications 1999

51. V. Isler, S. Kannan, S. Khanna, Randomized pursuit-evasion with local visibility. SIAM J. Discret. Math. **20**, 26–41 (2006)
52. J. Jonasson, O. Schramm, On the cover time of planar graphs. Electron. Comm. Probab. **5**, 85–90 (2000), (electronic)
53. S. Koenig, Y. Liu. Terrain coverage with ant robots: a simulation study, in *AGENTS'01* (2001)
54. S. Koenig, B. Szymanski, Y. Liu, Efficient and inefficient ant coverage methods. Ann. Math. Artif. Intell. **31**, 41–76 (2001)
55. S. Koenig, M. Likhachev, X. Sun. Speeding up moving-target search, in *AAMAS '07: Proceedings of the 6th International Joint Conference on Autonomous Agents and Multiagent Systems* (ACM, New York, NY, USA, 2007), pp. 1–8
56. B. Koopman, *Search and Screening: General Principles with Historical Applications* (Pergamon Press, Oxford, 1980)
57. R. Korf, Real-time heuristic search. Artifi. Intell. **42**, 189–211 (1990)
58. S. Kraus, O. Shehory, G. Taase, Coalition formation with uncertain heterogeneous information, in *Proceedings of the Second International Joint Conference on Autonomous Agents and Multiagent Systems* (2003), pp. 1–8
59. P. Lanillos, S.K. Gan, E. Besada-Portas, G. Pajares, S. Sukkarieh, Multi-uav target search using decentralized gradient-based negotiation with expected observation. Inf. Sci. **282**, 92–110 (2014)
60. S.M. LaValle, D. Lin, L.J. Guibas, J.C. Latombe, R. Motwani, Finding an unpredictable target in a workspace with obstacles, in *Proceedings of the 1997 IEEE International Conference on Robotics and Automation (ICRA-97)* (1997), pp. 737–742
61. Y.-Y. Liu, J.C Nacher, T. Ochiai, M. Martino, Y. Altshuler, Prospect theory for online financial trading. PloS one **9**(10), e109458 (2014)
62. L. Lovasz. Random walks on graphs: a survey. Combinatorics, Paul Erdos is Eighty, **2**, 353–398 (1996)
63. D. MacKenzie, R. Arkin, J. Cameron, Multiagent mission specification and execution. Auton. Robots **4**(1), 29–52 (1997)
64. E. Manisterski, R. Lin, S. Kraus, Understanding how people design trading agents over time, in *AAMAS '08: Proceedings of the 7th International Joint Conference on Autonomous Agents and Multiagent Systems* (International Foundation for Autonomous Agents and Multiagent Systems, Richland, SC, 2008), pp. 1593–1596
65. S. Mastellone, D.M. Stipanovi, C.R. Graunke, K.A. Intlekofer, M.W. Spong, Formation control and collision avoidance for multi-agent non-holonomic systems: theory and experiments. Int. J. Robot. Res. **27**(1), 107–126 (2008)
66. M.J. Mataric. *Interaction and Intelligent Behavior*. Ph.D. thesis, Massachusetts Institute of Technology (1994)
67. T.G. McGee, J. Karl Hedrick, Guaranteed strategies to search for mobile evaders in the plane, in *Proceedings of the 2006 American Control Conference* (2006), pp. 14–16
68. P.M. Morse, G.E. Kimball, *Methods of Operations Research* (MIT Press, Wiley, New York, 1951)
69. R. Olfati-Saber, Flocking for multi-agent dynamic systems: algorithms and theory. IEEE Trans. Autom. Control **51**(3), 401–420 (2006)
70. W. Pan, Y. Altshuler, A. Pentland, Decoding social influence and the wisdom of the crowd in financial trading network, in *2012 International Conference on Privacy, Security, Risk and Trust (PASSAT), and 2012 International Confernece on Social Computing (SocialCom)* (IEEE, 2012), pp. 203–209
71. L.E. Parker, C. Touzet, Multi-robot learning in a cooperative observation task. Distrib. Auton. Robot. Syst. **4**, 391–401 (2000)
72. M. Pfingsthorn, B. Slamet, A. Visser, A scalable hybrid multi-robot slam method for highly detailed maps, *RoboCup 2007: Robot Soccer World Cup XI*, vol. 5001, Lecture Notes in Computer Science (Springer, Berlin, 2008), pp. 457–464
73. S. Premvuti, S. Yuta, Consideration on the cooperation of multiple autonomous mobile robots, in *Proceedings of the IEEE International Workshop of Intelligent Robots and Systems* (1990), pp. 59–63

74. R. Puzis, Y. Altshuler, Y. Elovici, S. Bekhor, Y. Shiftan, A.S. Pentland, Augmented betweenness centrality for environmentally-aware traffic monitoring in transportation networks
75. M. Rehak, M. Pechoucek, P. Celeda, V. Krmicek, M. Grill, K. Bartos, Multi-agent approach to network intrusion detection, in *AAMAS '08: Proceedings of the 7th International Joint Conference on Autonomous Agents and Multiagent Systems* (International Foundation for Autonomous Agents and Multiagent Systems, Richland, SC, 2008) pp. 1695–1696
76. I. Rekleitis, V. Lee-Shuey, A. Peng Newz, H. Choset, Limited communication, multi-robot team based coverage, in *IEEE International Conference on Robotics and Automation* (2004)
77. I.M. Rekleitis, G. Dudek, E. Milios, Experiments in free-space triangulation using cooperative localization, in *IEEE/RSJ/GI International Conference on Intelligent Robots and Systems (IROS)* (2003)
78. S. Sariel, T. Balch, Real time auction based allocation of tasks for multi-robot exploration problem in dynamic environments, in *Proceedings of the AAAI-05 Workshop on Integrating Planning into Scheduling* (2005), pp. 27–33
79. R. Sawhney, K.M. Krishna, K. Srinathan, M. Mohan, On reduced time fault tolerant paths for multiple uavs covering a hostile terrain, in *AAMAS '08: Proceedings of the 7th International Joint Conference on Autonomous Agents and Multiagent Systems* (International Foundation for Autonomous Agents and Multiagent Systems, Richland, SC, 2008) pp. 1171–1174
80. J. Svennebring, S. Koenig, Building terrain-covering ant robots: a feasibility study. Auton. Robots **16**(3), 313–332 (2004)
81. B. Szymanski, S. Koenig, *The complexity of node counting on undirected graphs* (Technical Report, Computer Science Department, Rensselaer Polytechnic Institute, Troy, NY, 1998)
82. A. Thorndike, Summary of antisubmarine warfare operations in world war ii. Summary report, NDRC Summary Report (1946)
83. B.T. Sebastian, Efficient exploration in reinforcement learning — technical report cmu-cs-92-102. Technical report, Carnegie Mellon University (1992)
84. P. Vincent, I. Rubin, A framework and analysis for cooperative search using uav swarms, in *ACM Simposium on Applied Computing* (2004)
85. I.A. Wagner, Y. Altshuler, V. Yanovski, A.M. Bruckstein, Cooperative cleaners: a study in ant robotics. Int. J. Robot. Res. (IJRR) **27**(1), 127–151 (2008)
86. I.A. Wagner, A.M. Bruckstein, From ants to a(ge)nts: a special issue on ant robotics. Ann. Math. Artif. Intell. Special Issue on Ant Robot. **31**(1–4), 1–6 (2001)
87. I.A. Wagner, M. Lindenbaum, A.M. Bruckstein, Efficiently searching a graph by a smell-oriented vertex process. Ann. Math. Artif. Intell. **24**, 211–223 (1998)
88. H. Work, E. Chown, T. Hermans, J. Butterfield, Robust team-play in highly uncertain environments, in *AAMAS '08: Proceedings of the 7th International Joint Conference on Autonomous Agents and Multiagent Systems* (International Foundation for Autonomous Agents and Multiagent Systems, Richland, SC, 2008), pp. 1199–1202
89. A. Yamashita, M. Fukuchi, J. Ota, T. Arai, H. Asama, Motion planning for cooperative transportation of a large object by multiple mobile robots in a 3d environment, in *Proceedings of IEEE International Conference on Robotics and Automation* (2000), pp. 3144–3151
90. X. Zheng, S. Koenig, Robot coverage of terrain with non-uniform traversability, in *Proceedings of the IEEE International Conference on Intelligent Robots and Systems (IROS)* (2007), pp. 3757–3764
91. X. Zheng, S. Koenig, Reaction functions for task allocation to cooperative agents, in *Proceedings of the 7th International Joint Conference on Autonomous Agents and Multiagent Systems* (2008), pp. 559–566
92. X. Zheng, S. Jain, S. Koenig, D. Kempe, Multi-robot forest coverage, in *Proceedings of IROS* (press, 2005), pp. 3852–3857

Collaborative Patrolling Swarms
in Stochastically Expanding Environments

1 Introduction

The main problem that we discuss in this work is the *Dynamic Stochastic Cooperative Cleaners* problem — a problem assuming a regular grid of connected "squares" (or "pixels"), some of which are 'dirty' (such that the 'dirty' pixels form a connected region of the grid). On this dirty grid region several robotic agents operate, each having the ability to 'clean' the 'pixel' it is located in. We further assume that this connected 'dirty region' is stochastically evolving — where changes may take place, probabilistically, that are independent of, and certainly not caused by, the agents' activity. In the spirit of [23] we consider simple robots with only a bounded amount of memory (i.e. *finite-state-machines*).

The static variant of this problem was introduced in [47], where a cleaning protocol ensuring that a decentralized group of agents will jointly clean any given (and a-priori unknown) dirty region. The protocol's performance, in terms of cleaning time was fully analyzed and also demonstrated experimentally.

A dynamic generalization of the problem was later presented in [7], in which a deterministic expansion of the environment is assumed, simulating the spreading of a *contamination* (or a spreading "danger" zone or *fire*). Once again, the goal of the agents is to clean the spreading contamination in as efficiently as possible.

Various geometric properties of the problem and their connection to the minimal resources required in order to guarantee a successful completion of the task were analyzed in works such as [4–6, 11].

In this work we modify the 'dirty' region expansion model and add stochastic features to the spreadings of the region's pixels. We formally define and analyze the *Cooperative Cleaning* problem, this time under the stochastic dynamic generalization of it.

We concentrate in a variant of the *Cooperative Cleaning* problem, where the tiles have some probability to be contaminated by their neighbors' contamination. This

---

This chapter is based on work previously published in [41].

© Springer International Publishing AG 2018
Y. Altshuler et al., *Swarms and Network Intelligence in Search*,
Studies in Computational Intelligence 729, DOI 10.1007/978-3-319-63604-7_6

version of the problem has applications in the "real" world and in the computer network environments as well. For instance, one of the applications can be a distributed anti-virus software trying to overcome an epidemic malicious software attacking a network of computers. In this case, each infected computer has some probability to infect computers connected to it.

A more general paradigm of the cleaning problem is when the transformation of the contaminated area from one state to another is described as a function. For instance, following the previous example, we can say that the sub-network affected by the virus, is spreading by a certain rule. We can say that a computer will be infected by a virus with a certain probability, which depends on the number of the neighboring computers already infected. By defining rules for the contamination's spreading and cleaning, we can think of to this problem as a kind of Conway's "*Game of Life*", where each cell in the game's grid spreads its "seed" to neighboring cells (or alternatively, "dies") according to some basic rules.

While the problem posed in [7], as well as the analysis methods used and the correctness proofs, were all deterministic, it is interesting to examine the stochastic variant of such algorithms. In this work we analyze and derive a lower bound on the expected cleaning time for k agents running a cleaning protocol under a model, where every pixel, or "tile" in the neighborhood of the affected region may become contaminated at every time step with some probability, the contamination coming from its dirty neighbors.

The main results that are discussed in this work are a bound on the contaminated region's size and a bound on the cleaning time of the drones swarm, as well as a method which bounds the cleaning time for a given desired probability (presented in Sect. 4). We then provide an impossibility results for the problem, presented in Sect. 6.

2 Related Work

Significant research effort is invested in the design and simulation of multi-agent robotics and intelligent swarm systems (see e.g. [17, 25, 26, 36, 48, 49]).

In general, most of the techniques used for the distributed coverage of some region are based on some sort of cellular decomposition. For example, in [42] the area to be covered is divided between the agents based on their relative locations. In [24] a different decomposition method is being used, which is analytically shown to guarantee a complete coverage of the area. Reference [1] discusses two methods for cooperative coverage (one probabilistic and the other based on an exact cellular decomposition).

While some existing works concerning distributed (and decentralized) coverage present analytic proofs for the ability of the system to complete the task (for example, in [1, 18, 24]), most of them lack analytic bounds for the coverage time (and often extensive amounts of empirical results on this are made available by extensive simulations). Although a proof for the coverage completion is an essential element in the design of a multi-agent system, analytic indicators for its efficiency are in our opinion of great importance. We provide some such results, as bounds for the cleaning time of the agents, in Sect. 4.

An interesting work to mention in this context is that of Koenig and his collaborators [34, 43], where a swarm of ant-like robots is used for repeatedly covering an unknown area, using a real time search method called *node counting*. By using this method, the robots are shown to be able to efficiently perform a coverage mission, and analytic bounds for the coverage time are discussed.

Another work discussing a decentralized coverage of terrains is presented in [51]. This work examines domains with non-uniform traversability. Completion times are given for the proposed algorithm, which is a generalization of the forest search algorithm. In this work, though, the region to be searched is assumed to be known in advance - a crucial assumption for the search algorithm, which relies on a cell-decomposition procedure.

Vertex-Ant-Walk, a variation of the node counting algorithm is presented in [49] and is shown to achieve a coverage time of $O(n\delta_G)$, where δ_G is the graph's diameter, which is based on a previous work in which a cover time of $O(n^2\delta_G)$ was demonstrated [44]. Another work called *Exploration as Graph Construction*, provides a coverage of degree bounded graphs in $O(n^2)$ time, is described in [27]. Here a group of ant robots with a limited capability explores an unknown graph using special "markers". Another mechanism of using "pheromone markers" for the purpose of intra-swarm synchronization is discussed in [28], whereas a work that discusses a stochastic collaborative exploration of a virtual network for malicious cyber threats is discussed in [14]. Similar works concerning multi agents systems may be found in [1, 13, 18, 19, 24, 33, 37–39, 42]).

The *Cooperative Cleaning* problem is also strongly related to the problem of distributed search after mobile and evading target(s) [10, 15, 16, 22, 35, 40] or the problems discussed under the names of "Cops and Robbers" or "Lions and Men" pursuits [2, 20, 29–32].

3 Definitions

In this work we follow similar definitions that were used in the previous chapters, and specifically:

Definition 1 Let an undirected graph $G(V, E)$ describe the two dimensional integer grid $\mathbf{Z}^2$, whose vertices (or "*tiles*") have a binary property called "*contamination*". Let $cont_t(v)$ denote the contamination state of the tile v at time t, taking either the value "*on*" (for "dirty" or "contaminated") or "*off*" (for "clean").

For two vertices $v, u \in V$, the edge (v, u) may belong to E at time t only if both of the following hold:

1. v and u are $4-Neighbors$ in G.
2. $cont_t(v) = cont_t(u) = on$.

This however is a necessary but not a sufficient condition as we elaborate below.

The edges of E represent the connectivity of the contaminated region. At $t = 0$ all the contaminated tiles are connected, namely:

$$(v, u) \in E_0 \iff (v, u \text{ are } 4-Neighbors \text{ in } G) \wedge (cont_0(v) = cont_0(u) = on)$$

Edges may be added to E only as a result of a contamination spread and can be removed only while contaminated tiles are cleaned by the agents.

Definition 2 Let $F_t(V_{F_t}, E_t)$ be the contaminated sub-graph of G at time t, i.e.:

$$V_{F_t} = \{v \in G \mid cont_t(v) = on\}$$

We assume that F_0 is a single simply-connected component (the actions of the agents will be so designed that this property will be preserved).

Definition 3 Let ∂F denote the boundary of F. A tile is on the boundary if and only if at least one of its $8-Neighbors$ is not in F, meaning:

$$\partial F = \{v \mid v \in F \wedge 8-Neighbors(v) \cap (G \setminus F) \neq \emptyset\}$$

Definition 4 Let S_t denote the size of the dirty region F at time t, namely the number of grid points (or tiles) in F_t.

Let a group of k agents that can move on the grid G (moving from a tile to its neighbor in one time step) be placed at time t_0 on F_0, at some point $p_0 \in V_{F_t}$.

Definition 5 Let us denote by ΔF_t the *potential boundary*, which is the maximal number of tiles which might be added to F_t by spreading all the tiles of ∂F_t.

$$\Delta F_t \equiv \{v : \exists u \in \partial F_t \text{ and } v \in 4-Neighbors(u) \text{ and } v \notin F_t\}$$

As we are interested in the stochastic generalization of the dynamic cooperative cleaners model, we will assume that each tile in ΔF_t might be contaminated with some probability p. In the model we will analyze later, we assume that the status variables of the tiles of ΔF_t are independent from one another, and between time steps.

Definition 6 Let us denote by $\Phi_n(v)$ the *surrounding neighborhood* of a tile v, as the set of all the reachable tiles u from v within n steps on the grid (namely, the "digital sphere" or radius n around v). In this work we assume 4-connectivity among the region cells — namely, two tiles are considered as neighbors within one step iff the *Manhattan distance* between them is exactly 1.

The spreading policy, Ξ (v, ϕ, t), controls the contamination status of v at time $t + 1$, as a function of the contamination status of its neighbors in its nth digital sphere, at time t. Notice the Ξ (v, ϕ, t) can be also non deterministic.

Definition 7 Let us denote by Ξ (v, ϕ, t) the *spreading policy* of v as follows:

$$\Xi \ (v, \phi, t) : (V, \ \{\text{On, Off}\}^{\alpha}, \ \mathbb{N}) \to \{\text{On, Off}\}$$

where $\alpha \equiv |\Phi_n \ (v)|$, V denotes the vertices of the grid, $\{\text{On, Off}\}^{\alpha}$ denotes the contamination status of the members of $\Phi_n \ (v)$, at time t (for $t \in N$).

A basic example of using the previous definition of Ξ () is the case of the *deterministic model*, where at every d time-steps the contamination spreads from all tiles in ∂F_t to all the tiles in δF_t. This model can be defined using the Ξ function, as follows:

For every tile v we first define $\Phi_n \ (v)$ where n equals to 1 and assuming 4-connectivity. Then Ξ (v, ϕ, t), for any time-step t will be defined as follows:

$$\Xi \ (v, \phi, t) = \begin{cases} \text{On} & \text{if } t \mod d = 0 \text{ and } v \in \Delta F_t \\ \text{Off} & \text{Otherwise} \end{cases}$$

Notice that due to the fact that we are assuming that v is in ΔF_t, its *surrounding neighborhood* contains at least one tile with contamination status of *On*.

An interesting particular case of the general Ξ() function is the *simple uniform probabilistic spread*. In this scenario, a tile in $V \in \Delta F_t$ becomes contaminated with some predefined probability p, if and only if at least one of its nth neighbors are contaminated at time step t. Using the Ξ() function and the probability p, this can be formalized as follows:

For every tile v we first define $\Phi_n \ (v)$ where n equals to 1 and assuming 4-connectivity.

$$\Xi \ (v, \phi, t) = \begin{cases} \text{On} & \text{with probability of } p \text{ if } v \in \Delta F_t \\ \text{Off} & \text{Otherwise} \end{cases}$$

This model can naturally also be defined simply as:

$$\forall t \in \mathbb{N}, \forall v \in \Delta F_t, \ Prob \ (cont_{t+1}(v) = On) = p$$

In addition, we shall assume that the time is discrete.

In our work we will focus in this model, while deriving the analytic bounds for the cleaning time.

4 Lower Bound

4.1 Direct Bound

In this section we form a lower bound on the cleaning time of any cleaning protocol preformed by k agents. We start by setting a bound on the contaminated region's size at each time step, S_t. As we are interested in minimizing the cleaning time we should also minimize the contaminated region's area. Therefore we are interested in the minimal size of it, which achieved when the region's shape is sphere [7]. In our model each tile in the *potential boundary*, ΔF_t, has the same probability p to be contaminated in the next time step. The whole probabilistic process in each time step is *Binomial Distributed*, under the assumption that the spreading of each tile at any time step is independent from the spreadings of other tiles or from the spreadings of the same tile at different time steps.

As we are interested in the lower bound of the contaminated region's size we will assume that the expected number of newly added tiles to the contaminated region is minimal, which occurs when the region's shape forms a digital sphere (as presented in [8]). Then we can compute the expectation of this process for a specific time step t. Therefore, the size of the *potential boundary* is $\Delta F_t = 2\sqrt{2 \cdot S_t - 1}$ as shown in [8, 45].

Definition 8 Let us denote by X_t the random variable of the actual number of added tiles to the contaminated region at time step t.

Assuming the independence of tiles' contamination spreadings and given S_t, X_t is *Binomial Distributed*, $X_t|S_t \sim B(\Delta F_t, p)$, where each tile in the *potential boundary* has the same probability p to be contaminated. Therefore, we can say that the expectation of X_t given S_t is $\mu = E(X_t|S_t) = p \cdot \Delta F_t$.

Notice that occasionally the number of new tiles added to the contaminated region may be below μ. As we are interested in a lower bound, we should take some $\mu' < \mu$ such that: $Pr[X_t < \mu'|S_t] < \epsilon$, meaning that the probability that the number of the newly added tiles to the contaminated region is smaller than μ' is extremely small (tends to 0).

In order to bound X_t by some μ' we shall use the *Chernoff Bound*, where δ is the desired distance from the expectation, as follows:

$$Pr[X_t < (1 - \delta)\mu|S_t] < e^{-\frac{\delta^2 \mu}{2}}$$

Definition 9 Let us denote by q_t the probability that at time step t, the size of the added tiles to the contaminated region is not lower than $\mu' = (1 - \delta)\mu$ and it can be written as follows:

$$q_t = (1 - Pr[X_t < (1 - \delta)\mu|S_t])$$

Theorem 1 *Using any cleaning protocol, the area of the contaminated region at time step t can be recursively lower bounded, as follows:*

$$Pr\left[S_{t+1} \geq S_t - k + \left\lfloor 2 \cdot (1-\delta)\, p \cdot \sqrt{2 \cdot (S_t - k) - 1} \right\rfloor \Big| S_t \right] \geq q_t$$

Proof Notice that a lower bound for the contaminated region's size can be obtained by assuming that the agents are working with maximal efficiency, meaning that each time step every agent cleans exactly one tile.

In each step, the agents clean another portion of k tiles, but the remaining contaminated tiles spread their contamination to their $4-Neighbors$ and cause new tiles to be contaminated.

Lets us denote by the random variable S_{t+1} the number of contaminated tiles in the next time step. Using Definition 8 we can express S_{t+1} as follows:

$$S_{t+1} = S_t - k + X_t$$

Lets first bound the number of the added tiles using the *Chernoff Bound*. As X_t given S_t is *Binomial Distributed*, $X_t|S_t \sim B(\Delta F_t, p)$ and $\mu = E(X_t|S_t) = p \cdot \Delta F_t$. Using *Chernoff Bound* we know that:

$$Pr\left[X_t < (1-\delta)\mu|S_t\right] < e^{-\frac{\delta^2 \mu}{2}} \Rightarrow Pr\left[X_t < (1-\delta)p \cdot \Delta F_t|S_t\right] < e^{-\frac{\delta^2 \cdot p \cdot \Delta F_t}{2}}$$

Assigning $X_t = S_{t+1} - S_t + k$ from former definition of S_{t+1}, we get:

$$Pr\left[S_{t+1} - S_t + k < (1-\delta)p \cdot \Delta F_t|S_t\right] < e^{-\frac{\delta^2 \cdot p \cdot \Delta F_t}{2}}$$

As we are interested in the *minimal* number of tiles which can become contaminated at this stage. The minimal number of $4-Neighbors$ of any number of tiles is achieved when the tiles are organized in the shape of a "digital sphere" (see [8, 45]) - i.e. the *potential boundary* is $\Delta F_t = 2\sqrt{2 \cdot (S_t - k)} - 1$. Assigning ΔF_t value:

$$Pr\left[S_{t+1} < S_t - k + (1-\delta)p \cdot 2\sqrt{2 \cdot (S_t - k)} - 1|S_t\right] < e^{-\frac{\delta^2 \cdot p \cdot 2\sqrt{2 \cdot (S_t - k)} - 1}{2}}$$

As we are interested in the complementary event and using Definition 9

$$Pr\left[S_{t+1} \geq S_t - k + (1-\delta)p \cdot 2\sqrt{2 \cdot (S_t - k)} - 1|S_t\right] \geq 1 - e^{-\frac{\delta^2 \cdot p \cdot 2\sqrt{2 \cdot (S_t - k)} - 1}{2}} = q_t \tag{1}$$

As the number of tiles must be an integer value, we use $\left\lfloor (1-\delta) \cdot p \cdot 2\sqrt{2 \cdot S_t} - 1 \right\rfloor$ to be on the safe side. Using inequality Eq. 1 we get:

$$Pr\left[S_{t+1} \geq S_t - k + \left\lfloor 2 \cdot (1-\delta)\, p \cdot \sqrt{2 \cdot (S_t - k) - 1} \right\rfloor \Big| S_t \right] \geq q_t$$

Fig. 1 An illustration of the bound presented in Theorem 1. A lower bound for the contaminated region S_t, the area at time t, for various values of δ. The deterministic model (the Zig-Zag line) with ($d = 3$) compare to the stochastic model with $p = 1/3$ and $\delta \in [0.01, 0.1, 0.2, 0.3, 0.5]$ where both models have $k = 150$ and start with $S_0 = 20000$

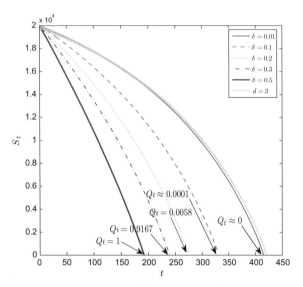

Figures 1 and 2 demonstrates the bound presented in Theorem 1. Note that as δ decreases the produced bound for the stochastic model is closer to the bound of the deterministic model, for $d = \frac{1}{p}$.

Definition 10 Let us denote by Q_t the *bound probability* that the contaminated region's size at time step t will be [at least] S_t. Q_t can be expressed as follows:

$$Q_t = \prod_{i=0}^{t} q_i$$

Notice that for bounding the area of the region at time step t using Theorem 1 and Definition 10, the bound, which will be achieved, will be in probability Q_t. We want to have Q_t sufficiently high.

We shall assume, for the sake of analysis, that the dynamic value of the area, S_t, is always kept not less than some $\hat{S} < S_0 - k + \lfloor 2 \cdot (1 - \delta) p \cdot \sqrt{2 \cdot (S_0 - k) - 1} \rfloor$ (as we want S_1 to be bigger or equal to $\hat{S}$). Then the next Lemma holds:

Lemma 1 *For any $T \geq 1$, if for all $1 \leq t \leq T$ the contaminated region's size S_t is always kept not less than some $\hat{S} < S_0 - k + \lfloor 2 \cdot (1 - \delta) p \cdot \sqrt{2 \cdot (S_0 - k) - 1} \rfloor$ then:*

$$Q_T \geq \hat{Q}_T = \hat{q}^T = \left(1 - e^{-\delta^2 \cdot p \cdot 2\sqrt{2 \cdot (\hat{S} - k) - 1}}\right)^T$$

Proof We will prove this Lemma by induction on T.

Fig. 2 An illustration of the bound presented in Theorem 1. A lower bound for the contaminated region S_t, the area at time t, for various number of agents k. The lower bound for the problem parameters $S_0 = 20000$, $p = 0.5$ and $\delta = 0.3$ for different number of agents k

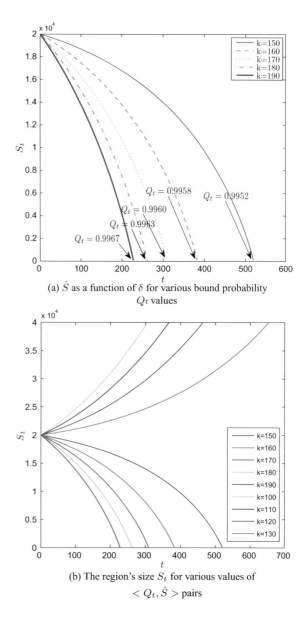

(a) $\hat{S}$ as a function of δ for various bound probability Q_t values

(b) The region's size S_t for various values of $< Q_t, \hat{S} >$ pairs

- *Base step:* For $T = 1$, we can write the probabilities Q_1 and $\hat{Q}_1$ as follows:

$$Q_1 = q_1 \geq 1 - e^{-\delta^2 \cdot p \cdot 2\sqrt{2 \cdot (S_1 - k) - 1}}$$

$$\hat{Q}_1 = \hat{q} \geq 1 - e^{-\delta^2 \cdot p \cdot 2\sqrt{2 \cdot (\hat{S} - k) - 1}}$$

As we assume that $S_t \geq \hat{S}$ than it is not hard to see that $Q_1 \geq \hat{Q}_1$.

- *Induction hypothesis:* We assume that for all $T \leq T'$ holds $Q_T \geq \hat{Q}_T$.
- *Induction step:* We will prove that for $T = T' + 1$ holds $Q_T \geq \hat{Q}_T$. From Definition 10 we can write the probability Q_T as follows:

$$Q_T = Q_{T'+1} = Q_{T'} \cdot q_{T'+1}$$

By the induction hypothesis we know that for all $T \leq T'$ holds $Q_T \geq \hat{Q}_T$ than we can rewrite $Q_{T'+1}$ and also writing $\hat{Q}_T$ for $T = T' + 1$ we get that:

$$Q_T = Q_{T'+1} \geq \hat{Q}_{T'} \cdot q_{T'+1}$$

$$\hat{Q}_T = \hat{Q}_{T'+1} = \hat{Q}_{T'} \cdot \hat{q}$$

As we want to compare these probabilities and to prove that $Q_T \geq \hat{Q}_T$ all we need to show is that $q_{T'+1} \geq \hat{q}$. As we assume that for all $1 \leq t \leq T' + 1$ holds that $S_t \geq \hat{S}$ and particularly for $t = T' + 1$, than it is not hard to see that $q_{T'+1} \geq \hat{q}$ and therefore $Q_T \geq \hat{Q}_T$.

Theorem 2 *For any contaminated region of size S_0, using any cleaning protocol, the probability that $S_{\hat{\tau}_\delta}$, the contaminated area at time step $t = \hat{\tau}_\delta$, is greater or equal to some $\hat{S} < S_0 - k + \lfloor 2 \cdot (1 - \delta) \, p \cdot \sqrt{2 \cdot (S_0 - k) - 1} \rfloor$ can be lower bounded, as follows:*

$$Pr\left[S_{\hat{\tau}_\delta} \geq \hat{S} \right] \geq \left(1 - e^{-\delta^2 \cdot p \cdot 2\sqrt{2 \cdot (\hat{S} - k) - 1}} \right)^{\hat{\tau}_\delta}$$

where:

$$\hat{\tau}_\delta \triangleq \frac{\sqrt{\varpi \cdot \left(\hat{S} - k - \frac{1}{2} \right)} - \sqrt{\varpi \cdot \left(S_0 - k - \frac{1}{2} \right)} + \ln \left(\frac{\sqrt{\varpi \cdot \left(\hat{S} - k - \frac{1}{2} \right)} - \frac{k}{2}}{\sqrt{\varpi \cdot (S_0 - k - \frac{1}{2})} - \frac{k}{2}} \right)^{\frac{k}{2}}}{\varpi}$$

and

$$\varpi \triangleq 2(1 - \delta)^2 \cdot p^2$$

Proof Observe that by denoting $y_t \triangleq S_t$ Theorem 1 can be written as:

$$y_{t+1} - y_t \geq \left\lfloor 2 \cdot (1 - \delta) \cdot p\sqrt{2 \cdot (y_t - k) - 1} \right\rfloor - k$$

Searching for the minimal area we can look at the equation:

$$y_{t+1} - y_t = \left\lfloor 2 \cdot (1 - \delta) \cdot p\sqrt{2 \cdot (y_t - k) - 1} \right\rfloor - k$$

By dividing both sides by $\Delta t = 1$ we obtain:

$$y_{t+1} - y_t \triangleq y' = \left\lfloor \sqrt{(1-\delta)^2 \cdot p^2 \cdot 8 \left[y - \left(k + \frac{1}{2} \right) \right]} \right\rfloor - k \qquad (2)$$

Notice that the values of y', the derivative of the change in the region's size, might be positive (stating an increase in the area), negative (stating a decrease in the area), or complex numbers (stating that the area is smaller than k, and will therefore be cleaned before the next time step).

Let us denote $x^2 \triangleq (1-\delta)^2 \cdot p^2 \cdot 8 \left[y - \left(k + \frac{1}{2} \right) \right]$. After calculating the derivative of both sides of this expression we see that:

$$2x \cdot x' = (1-\delta)^2 \cdot p^2 \cdot 8 y'$$

and after using the definition of y' of Eq. 2 we see that:

$$2x \cdot \frac{dx}{dt} = 2x \cdot x' = (1-\delta)^2 \cdot p^2 \cdot 8 \left(\left\lfloor \sqrt{(1-\delta)^2 \cdot p^2 \cdot 8 \left[y - \left(k + \frac{1}{2} \right) \right]} \right\rfloor - k \right)$$

$$\leq (1-\delta)^2 \cdot p^2 \cdot 8 \, (x - k) \qquad (3)$$

From Eq. 3 a definition of dt can be extracted:

$$dt \geq \frac{1}{8 \cdot (1-\delta)^2 \cdot p^2} \cdot \frac{2x}{x-k} dx$$

$$\geq \frac{1}{4(1-\delta)^2 \cdot p^2} \cdot \frac{x-k+k}{x-k} dx \geq \frac{1}{4(1-\delta)^2 \cdot p^2} \left(1 + \frac{k}{x-k} \right) dx$$

The value of x can be achieved by integrating the previous expression as follows (notice that we are interested in the equality of the two expressions):

$$\int_{t_0}^{t} dt = \int_{x_0}^{x} \frac{1}{4(1-\delta)^2 \cdot p^2} \left(1 + \frac{k}{x-k} \right) dx$$

After the integration we can see that:

$$i \Big|_{t_0}^{t} = \frac{1}{4(1-\delta)^2 \cdot p^2} (x + k \ln (x - k)) \Big|_{x_0}^{x}$$

and after assigning $t_0 = 0$:

$$4(1-\delta)^2 \cdot p^2 \cdot t = x - x_0 + k \ln \frac{x-k}{x_0 - k}$$

Returning back to y and using ϖ definition we get:

$$\varpi \cdot t = \sqrt{\varpi \left(y - k - \frac{1}{2} \right)} - \sqrt{\varpi \left(y_0 - k - \frac{1}{2} \right)} + \ln \left(\frac{\sqrt{\varpi \left(y - k - \frac{1}{2} \right)} - \frac{k}{2}}{\sqrt{\varpi \left(y_0 - k - \frac{1}{2} \right)} - \frac{k}{2}} \right)^{\frac{k}{2}}$$

Returning to the original size variable S_t, we see that:

$$\varpi \cdot t = \sqrt{\varpi \left(S_t - k - \frac{1}{2} \right)} - \sqrt{\varpi \left(S_0 - k - \frac{1}{2} \right)} + \ln \left(\frac{\sqrt{\varpi \left(S_t - k - \frac{1}{2} \right)} - \frac{k}{2}}{\sqrt{\varpi \left(S_0 - k - \frac{1}{2} \right)} - \frac{k}{2}} \right)^{\frac{k}{2}}$$

(4)

Defining that $\hat{\tau}_\delta = t$ and combining Eq. 4 with Lemma 1 knowing that $S_{t'} \geq \hat{S}$ for all $1 \leq t' \leq \hat{\tau}_\delta$ we get the following inequality:

$$Pr \left[S_{\hat{\tau}_\delta} \geq \hat{S} \right] \geq \left(1 - e^{-\delta^2 \cdot p \cdot 2 \sqrt{2 \cdot (\hat{S} - k) - 1}} \right)^{\hat{\tau}_\delta}$$

where:

$$\hat{\tau}_\delta \triangleq \frac{\sqrt{\varpi \cdot \left(\hat{S} - k - \frac{1}{2} \right)} - \sqrt{\varpi \cdot \left(S_0 - k - \frac{1}{2} \right)} + \ln \left(\frac{\sqrt{\varpi \cdot \left(\hat{S} - k - \frac{1}{2} \right)} - \frac{k}{2}}{\sqrt{\varpi \cdot (S_0 - k - \frac{1}{2})} - \frac{k}{2}} \right)^{\frac{k}{2}}}{\varpi}$$

and

$$\varpi \triangleq 2(1 - \delta)^2 \cdot p^2$$

In Theorem 2 we can guarantee with high probability of $\hat{Q}_{\tau_\delta}$ that the contamination region's size will not be lower than $\hat{S}$ - namely for any time step $t > \tau_\delta$ the probability $Pr[S_t \geq \hat{S}]$ is getting lower and therefore the probability that the agents will succeed in cleaning the contaminated area is increasing. We are showing that by choosing small enough $\hat{S}$ so we know that the agents will succeed in cleaning the rest of the 'dirty' region, we will be able to guarantee with high probability the whole cleaning of the 'dirty' region. For example, choosing $\hat{S}$ to be in $o(k)$ will assure that for $S_t \leq \hat{S} \leq c \cdot k$ for some small constant c, the rest of the contaminated region will be cleaned in at most c time steps by the k cleaning agents.

An illustration of the bound on the cleaning time, as presented in Theorem 2, is shown in Figs. 3 and 4 and the corresponding bound probability in Figs. 5 and 6. Notice that as $\hat{S}$ and δ increase the cleaning time decreases.

Fig. 3 An illustration of the bound presented in Theorem 2 of the cleaning time t in order to reach $\hat{S}$. The bound on the cleaning time where $p = 0.5$ and $\delta \in [0.27, 0.3, 0.4, 0.5, 0.6, 0.7]$ where the cleaning done by $k = 150$ agents and starting with $S_0 = 20000$

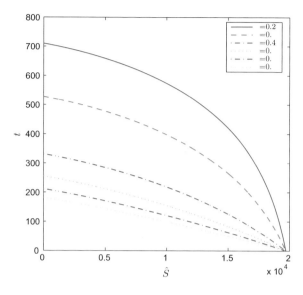

Fig. 4 An illustration of the bound presented in Theorem 2 of the cleaning time t in order to reach $\hat{S}$. The lower bound on the cleaning time for $S_0 = 20000$, $p = 0.5$ and $\delta = 0.3$ for different number of agents $k \in [150, 160, 170, 180, 190]$

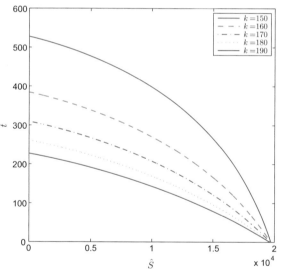

4.2   Using the Bound

In Theorem 2 we presented a bound which guarantees that the contaminated region's size will not be smaller than some predefined size $\hat{S}$ with the bound probability, Q_t. We suggest a method which make this bound useful when one willing to be guaranteed of successfully cleaning of the contaminated area with some desired probability Q_t in a certain model, where the initial contaminated region's size is S_0, each one of the

Fig. 5 An illustration of the
probability produced by the
bound presented in
Theorem 2. The bound
probability Q_t for initial
region's size $S_0 = 20000$,
spreading probability
$p = 0.5$ and number of
agents $k = 150$ for the
following values of
$\delta \in [0.4, 0.3, 0.25, 0.2]$

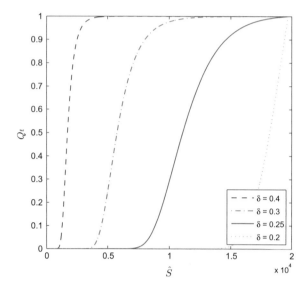

Fig. 5 An illustration of the
probability produced by the
bound presented in
Theorem 2. The bound
probability Q_t for initial
region's size $S_0 = 20000$,
spreading probability
$p = 0.5$ and number of
agents $k = 150$ for the
following values of
$\delta \in [0.4, 0.3, 0.25, 0.2]$

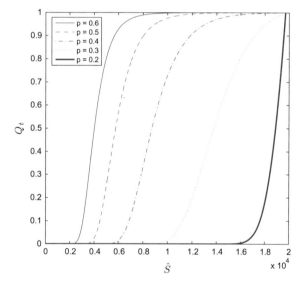

Fig. 6 An illustration of the
probability produced by the
bound presented in
Theorem 2. The bound
probability Q_t for
$S_0 = 20000$, $\delta = 0.3$ and
$k = 150$ for the following
values of
$p \in [0.6, 0.5, 0.4, 0.3]$

tiles in the surrounding neighborhood of the contaminated area has a probability p
to be contaminated by each one of its neighbors's contamination spreads and with k
cleaning agents.

Notice that the only free variables left in the bound are the analysis parameters δ
and $\hat{S}$. We should also notice the fact that as δ decreases the "usefulness" of the bound
decreases (see Fig. 6) because when δ is closer to 0, the model tern to the deterministic
variant of the *cooperative cleaning* problem. In this variant of the problem the bound,

as presented in Theorem 2, will "predict" that the contamination will spread exactly by the *potential boundary* mean in every step (as shown in Sect. 4.1). Furthermore, as δ increases, although the "usefulness" of the bound increases, the predicted bound is the naive one, where at each step there are no spreads and all the k agents clean perfectly - i.e. the size of the contaminated region at time step t will be exactly $S_t = S_0 - t \cdot k$ (which can be guaranteed in high probability).

We suggest the following method, in order to eliminate the need to identify the analysis parameters. Once someone willing to use this bound he should provide the desired bound probability - Q_t and the parameter of the model. Then for each value of δ in the range of [0, 1] he should find the corresponding value of the minimal $\hat{S}$ which satisfies the inequality $Pr\left[S_t \geq \hat{S}\right] \geq Q_t$ as illustrated in Fig. 7. As $\delta \in \mathbf{R}$ - i.e. a real number, once using this method we should choose the granularity of δ for which calculate the appropriate $\hat{S}$.

Notice that there can be exists some minimal value of $\delta = \delta_{MIN}$, where for any value of $\delta < \delta_{MIN}$ there is no solution for the bound inequality. Also notice that there can be exists some maximal value of $\delta = \delta_{MAX}$, where for any value of $\delta > \delta_{MIN}$ the corresponding $\hat{S}$ is the same as for δ_{MAX}.

Furthermore, for each pair of values of δ and $\hat{S}$ there exists its corresponding cleaning precess of k agents with initial region's size S_0 as demonstrated in Fig. 7b. Each curve bounds the cleaning process from S_0 to the applicable $\hat{S}$.

As we are interested in finding the tightest bound, looking at the frontier of the bounds, as shown in Fig. 7b, we can combine the relevant curves to one comprehensive bound. This bounds integrates the bound for a specific range of j values of $\delta \in \left[\delta_{i_1}, \delta_{i_2}, ..., \delta_{i_j}\right]$, where for each time step t we choose the maximal S_t as illustrated in Fig. 8.

Notice that the combined bound is independent of the selection of values for analysis parameters δ and $\hat{s}$. Also we can notice that this bound limits the contaminated region's size to some minimal $\hat{S}_{MIN}$ where we almost certain that the agents will succeed in terminating the cleaning process successfully.

Notice that the inequality in Theorem 2 bounds the probability that the contaminated region's size at time step t will not be smaller than $\hat{S}$, therefore, an increase in the spreadings probability causes to an increase in the expected contaminated region's size and thus increases the probability $\left(\hat{q}\right)^t$ (as demonstrated in Fig. 6).

Notice that as illustrated in Fig. 9, as the number of agents increases the probabilistic and the deterministic bounds are more similar. This result is not surprising considering the method we presented. In our method, in order to make the bound tighter, we favor the lower values of δ. As the number of agents increases the bound can be guaranteed in the desired probability with lower values of δ and as δ decreases the model becomes more similar to the deterministic one (Fig. 10).

Fig. 7 Figure **a** is an illustration of $\hat{S}$ as a function of δ using the bound presented in Theorem 2 for the following model parameters – $S_0 = 20000$, $p = 0.1$ and $k = 50$ for various values of Q_t. Figure **b** is an illustration of the cleaning process for various pairs of values of $\hat{S}$ and δ as shown in Figure **a** for the same model parameters with $Q_t = 0.95$

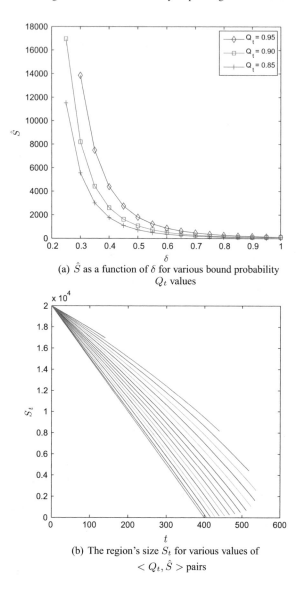

(a) $\hat{S}$ as a function of δ for various bound probability Q_t values

(b) The region's size S_t for various values of $< Q_t, \hat{S} >$ pairs

4.3 Parameters Selection

One of the problems of the bound as brought in Sect. 4.1 is the nature of the probabilistic bounds to decay to 0 (as shown in Figs. 5 and 6), which caused due to the fact that the bound probability Q_t is a product of each step's probability, q_t, and because as t increase Q_t decreases. One of the reasons which explains this problem is a bad selection of parameters - e.g. in the bound for the cleaning time (Eq. 4 a selection

Fig. 8 An illustration of the combined bound for the bounds as shown in Figs. 7a and 7b for the same model parameters – $S_0 = 20000$, $p = 0.1$ and $k = 50$ for various values of Q_t compared to the deterministic model with $d = 10$ and to the naive bound S_0/k

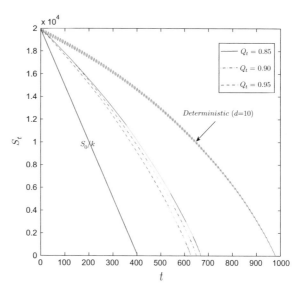

Fig. 9 A comparison between the deterministic bound and the probabilistic bound as a function of the desired guaranteeing probability - Q_t for various contaminated region's sizes and number of cleaning agents

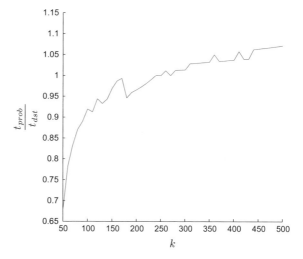

of too small $\hat{S}$ will lead to fast decay in the probability. Furthermore, there exist trade-offs, when selecting the parameters' values, between the bound results and the the probability which guarantees its likelihood (e.g. see Fig. 11).

One way to avoid this problem is by selecting $\hat{S}$ as big as possible, as in Figs. 5 and 11. As $\hat{S}$ increases the probability of each time step, q_t, increases and so the total probability Q_t. Another technique for eliminating this problem is by "wrapping" number of time steps into one, thus artificially decreasing the time and therefore decreasing the power of q_t in Q_t (in Lemma 1).

Fig. 10 Figures **b** and **a**
compare the deterministic
bound and the probabilistic
bound as a function of the
desired guaranteeing
probability - Q_t for various
contaminated region's sizes
and number of cleaning
agents

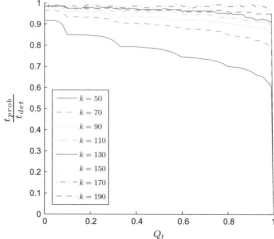

(a) The ratio between the bounds as function of the bound
probability Q_t for various number of agents k

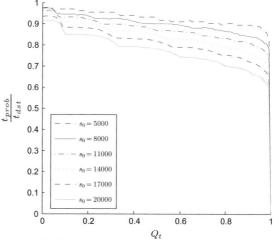

(b) The ratio between the bounds as function of the bound
probability Q_t for various values of initial region's size S_0

Another example of the trade-off in choosing the parameters can be shown in
Fig. 6 where Q_t is illustrated for various values of the probability p. Interestingly, as
p decreases our confidence in the bound result is decreasing although we know that
the the agents preforming the cleaning protocol have a better chance to successfully
complete their work.

Fig. 11 Parameters
selection. In **a** we can see Q_t
for $S_0 = 20000$ and $k = 150$
for the following of values of
$p \in [0.4, 0.3, 0.2, 0.1]$. In **b**
we can see Q_t for
$S_0 = 20000$ and $p = 0.4$ for
the following of values of
$k \in [150, 250, 350, 450,$
$550, 650]$

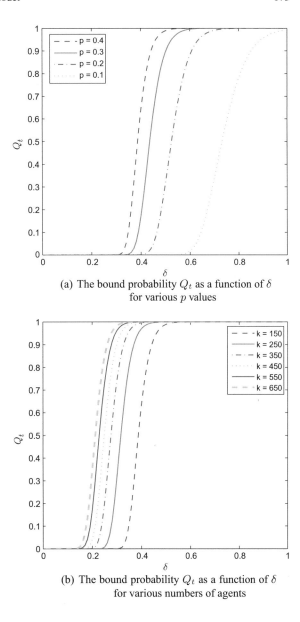

(a) The bound probability Q_t as a function of δ
for various p values

(b) The bound probability Q_t as a function of δ
for various numbers of agents

5   Lower Bound - Markovian Model

In Sects. 4.1 and 4.2 we presented probabilistic lower bounds on the contaminated
region's size and the cleaning time where in order to achieve the latter, we assumed
that the contaminated region's size is bounded from below by some $\hat{S}$. In this section
we will assume that the contaminated region is also bounded from above - i.e. the

area is bounded from above and below by some S^{MAX} and S^{MIN} respectively. The method presented in this section bounds the probability that as the time t tends to infinity the contaminated region will be clean - i.e. its size will be *at most* S^{MIN} or totally contaminated - i.e. its size will be *at least* S^{MAX}. This method can also be used to bound the probability that at a specific time step t the contaminated regions size will not be lower than some S.

We start by setting a bound on the contaminated region's size at each time step, S_t. As we are interested in minimizing the cleaning time we should also minimize the contaminated region's area. Therefore we are interested in the minimal size of it, which achieved when the region's shape is sphere [7]. In our model each tile in the *potential boundary*, ΔF_t, has the same probability p to be contaminated in the next time step. The whole probabilistic process in each time step is *Binomial Distributed*, under the assumption that the spreading of each tile at any time step is independent from the spreadings of other tiles or from the spreadings of the same tile at different time steps.

As we are interested in the lower bound of the contaminated region's size we will assume that the expected number of newly added tiles to the contaminated region is minimal, which occurs when the region's shape forms a "digital sphere" (as presented in [8]). Therefore, the size of the *potential boundary* is upper bounded by $\Delta F_t = 2\sqrt{2 \cdot S_t - 1}$ as shown in [8, 45] and lower bounded by $\Delta F_t = 0$. Then we can compute for a specific time step t, the exact probability that in the next time step the contaminated regions size will be in some size of j tiles.

The process of randomly choosing the number of added tiles to the contaminated region, and also the series of the contaminated region's area through the time, can be referred as a *Markov Chain* where the probability to change from one state to another depends only on the current state. In other words, at each time step the size of the contaminated area is depend only on the size of it in the previous time step. At each time step there is a finite number of options to add tiles to the contaminated region and for each one of them we can compute the probability to add it.

In this case, we can explicitly write the probability to change the contaminated region from one size to another.

Definition 11 Let us denote by S^i the state of the contaminated region when its size is i tiles.

Definition 12 Let us denote by $P_{i,j}$ the probability to change the contaminated region's size from state S^i to state S^j in one time step. And the probability can be written as:

$$P_{i,j} = Pr[S_t = S^j | S_{t-1} = S^i]$$

Definition 13 Let us denote by P_t^i the probability that the contaminated region's size will be i at time step t. And the probability can be write as:

$$P_t^i = Pr[S_t = S^i]$$

As presented in Sect. 4.1, assuming that the contaminated region shape is always kept as a "digital sphere", we can calculate the probability $P_{i,j}$ for any two states S^i and S^j. From each state S^i we have at most $2\sqrt{2(S^i - k)} - 1$ options to change the state, under the assumption of perfect cleaning work done by the k cleaning agents.

Theorem 3 *Regardless of the cleaning protocol, which performed by k cleaning agents, the probability $P_{i,j}$, to change the contaminated region's state S^i to state S^j for any two states, can be lower bounded as follows:*

$$P_{i,j} = \begin{cases} \binom{2\sqrt{2(i-k)-1}}{j-i+k} \cdot p^{j-i+k} \cdot (1-p)^{2\sqrt{2(i-k)-1}-(j-i+k)} & \text{if } S^j \in \mathbb{V} \\ 0 & \text{Otherwise} \end{cases}$$

where:

$$\mathbb{V} = \left[S^i - k .. S^i - k + 2\sqrt{2(S^i - k) - 1} \right]$$

Proof Notice that a lower bound for the contaminated region's size can be obtained by assuming that the agents are working with maximal efficiency, meaning that for each state S^i every agent cleans exactly one tile.

Therefore, in each state, the agents clean another portion of k tiles, but the remaining contaminated tiles spread their contamination to their $4 - Neighbors$ and cause new tiles to be contaminated.

As we are interested in the *minimal* number of tiles which can become contaminated at this stage. The minimal number of $4 - Neighbors$ of any number of tiles is achieved when the tiles are organized in the shape of a "digital sphere" (see [8, 12]) - i.e. the *potential boundary* is $\Delta F_t = 2\sqrt{2 \cdot (S_t - k)} - 1$.

Using Definition 8 of the random variable X_t of the newly added tiles to the contaminated region we can express the number of tiles when changing from one state S^i to another state S^j.

Then by choosing to contaminate $X_t = j - i + k$ tiles from the *potential boundary* with probability p for each tile, and not contaminating the rest of the tiles in the *potential boundary* with probability $(1 - p)$ - i.e. this process is binomial distributed and the rest of the proof is implied.

When changing from one state to another, we choose $j - i + k$ tile from the *potential boundary* to get contaminated with probability p. We also need to pay special attention to the lower margin of the contaminated region's states, where the size can decrease under the number of contaminated tiles in state s^{MIN}. In this scenario we will assume that $P_{i,MIN} = \sum_{j \leq MIN} P_{i,j}$.

Assuming also that the size of the contaminated region is bounded by some maximal state S^{MAX}, we should also consider the upper marginal states as, $P_{i,MAX} = \sum_{j \geq MAX} P_{i,j}$.

Notice that unlike the analysis performed in Sect. 4.1 (where the area of the contaminated region is lower bounded by zero, but is not upper bounded), in the following analysis we assume that this area is also upper bounded (by some very large predefined constant). This assumption is required in order to use a Markovian model.

Definition 14 Let us denote by $M_{k,p,S^{MIN},S^{MAX}}$ as the *Transition Matrix* of the cleaning problem using k agents, with bound on the contaminated region's area of S^{MAX} and S^{MIN} and a probability p of a tile in the *potential boundary* to get contaminated.

$$M_{k,p,S^{MIN},S^{MAX}} = \begin{pmatrix} P_{MIN,MIN} & P_{MIN,MIN+1} & \cdots & P_{MIN,MAX} \\ P_{MIN+1,MIN} & P_{MIN+1,MIN+1} & \cdots & P_{MIN+1,MAX} \\ \vdots & \vdots & \ddots & \vdots \\ P_{MAX,MIN} & P_{MAX,MIN+1} & \cdots & P_{MAX,MAX} \end{pmatrix}$$

Lemma 2 *The transition matrix -* $M_{k,p,S^{MIN},S^{MAX}}$ *is* irreducible *and* aperiodic.

Proof The transition matrix - $M_{k,p,S^{MIN},S^{MAX}}$ is an *irreducible* matrix if for every two states S^i and S^j there exists a non-zero probability of transitioning from state S^i to state S^j with $n \geq 0$ steps. From Theorem 3 and Definition 14 it is not hard to see that for each pair of states there is a non-zero probability of transitioning from one state to another and therefore $M_{k,p,S^{MIN},S^{MAX}}$ is an *irreducible* matrix. Also each state can be re-visited aperiodically - namely, the GCD of the recurrence in each state is $k = 1$, because there is some non-zero probability to stay in each state and hence $M_{k,p,S^{MIN},S^{MAX}}$ is also *aperiodic*.

Notice that we assumes that there are no states which it is impossible to get out from. Meaning that the number of the contaminated region of the minimal state S^{MIN} is larger than the number of cleaning agents k. Otherwise, for states S^i where $i \leq k$, the agents will clean all the contaminated tiles and there will be no spreads, which ends in a final state S^0 without any option to change this state anymore.

Lemma 3 *The transition matrix -* $M_{k,p,S^{MIN},S^{MAX}}$ *is a* stochastic matrix.

Proof From Theorem 3 and Defenition 14 we know that each row i is the probability to leave state S^i to any other state and this probability is *binomial distributed* and summed to 1. Therefor $M_{k,p,S^{MIN},S^{MAX}}$ is a *stochastic matrix*.

Theorem 4 *The* Perron–Frobenius Theorem *holds for the transition matrix -* $M_{k,p,S^{MIN},S^{MAX}}$.

Proof $M_{k,p,S^{MIN},S^{MAX}}$ is a non-negative and irreducible matrix and therefor The *Perron–Frobenius Theorem* holds for it.

Consequently from Theorem 4, $M_{k,p,S^{MIN},S^{MAX}}$ has a real simple *principle eigenvalue* λ_1, which is bigger than all the other eigenvalues i.e. $\forall i \neq 1 : |\lambda_1| > |\lambda_i|$. There is also a unique positive unit eigenvector corresponds to λ_1, denoted by V_{λ_1} which is the only eigenvector of $M_{k,p,S^{MIN},S^{MAX}}$ whose components are all positive (see Perron–Frobenius Theorem in [46]).

Also from Theorem 4, Lemma 3 and *The ergodic theorem* (see [21]) we know that the *principle eigenvector* V_{λ_1} is a unique distribution row vector and is the *stationary distribution* of the Markov Chain defined by the transition matrix $M_{k,p,S^{MIN},S^{MAX}}$. Furthermore, for any distribution row-vector q:

$$\lim_{t \to \infty} \left(M_{k,p,S^{MIN},S^{MAX}} \right)^t \cdot q = V_{\lambda_1}$$

In other word, the probability to be in each of the states when the time t tends to infinity is defined by the principle eigenvector V_{λ_1} regardless of the initial state. Using Definition 13 we can infer that:

$$\left(M_{k,p,S^{MIN},S^{MAX}} \right)^t = \begin{pmatrix} P_t^{MIN} \\ P_t^{MIN+1} \\ \vdots \\ P_t^{MAX} \end{pmatrix}$$

In order to compute the principle eigenvector of $M_{k,p,S^{MIN},S^{MAX}}$ and the probability distribution in a given time step t for a given starting state one can use *Power-Method*. The Power Method is an iterative algorithm to compute (under certain conditions) the eigenvalue of largest absolute value and the corresponding eigenvector of a general $N \times N$ matrix M. Using this method infinitely times (or t times) will lead to the desired probability distribution to succeed/fail in the cleaning process (for further reading, see for example [50]).

The results of this technique are demonstrated in Fig. 12, where we can notice that there is a short transition phase as the number of the cleaning agents increases.

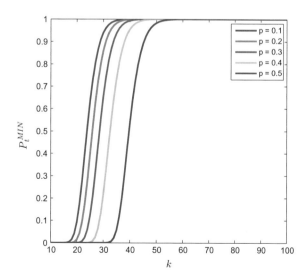

Fig. 12 An illustration of the success probability P_t^{MIN} using Theorem 4 as a function of the number of cleaning agents k and for various values of the spreading probability p. With $S^{MAX} = 5000$ and $S^{MIN} = O(k) > k$

Namely, as the number of cleaning agents increases the probability for a successful cleaning, P_t^{MIN}, increases and tends to 1. It is important to note though that even when the success probability P_t^{MIN} tends to 1, the successful cleaning is not always guaranteed, as this is a lower bound assuming a "perfect world" scenario.

It is also important to remember that using too small values for S^{MAX} will lead to an unbalanced stationary probability distribution – namely, resulting in only one possible final state (which is S^{MIN}), whereas we would prefer the stationary probability to be distributed between S^{MAX} and S^{MIN}.

6 Impossibility Results

6.1 Direct Result

While the theoretical lower bound presented in Sect. 4 can decrease the uncertainty regarding the existence of a solution for the cleaning problem with certain number of agents, one might be interested in the opposite question, namely — how can we guarantee that a group of drone agents *will not* be able to successfully accomplish the cleaning work (regardless of the cleaning protocol being used or the contaminated region's properties e.g. its shape and spreading probability). In contrary to the impossibility result in the deterministic model (as in [3, 13]), in the stochastic model of the *Cooperative Cleaning* problem, the probabilistic characteristics of the problem should be considered. Consequently, we will set the impossibility result with probabilistic restrictions as follows:

Theorem 5 *Using any cleaning protocol, k agents cleaning a contaminated region, where each tile in the* potential boundary *may be contaminate by already contaminated neighboring tiles with some probability p in each time step, will not be able to clean this contaminated region if:*

$$S_0 > \left\lfloor \frac{k^2}{8 \cdot p^2} + k + \frac{1}{2} \right\rfloor$$

with the probability Q_t for any time step t - i.e.:

$$\forall t \, Pr\,[S_t \geq S_0] = Q_t$$

Proof Firstly, we shall require that the contaminated region's size increases between each time step, guaranteeing us that the contaminated region's size will keep on growing, and thus impossible to be cleaned. Therefore we want that in each time step t the size of the contaminated region will be bigger than the previous one - i.e.:

$$S_{t+1} - S_t > 0$$

Using Theorem 1 we know that:

$$S_{t+1} \geq S_t - k + \left\lfloor 2 \cdot (1 - \delta) \, p \cdot \sqrt{2 \cdot (S_t - k) - 1} \right\rfloor$$

and therefore we shall require that:

$$\left\lfloor 2 \cdot (1 - \delta) \, p \cdot \sqrt{2 \cdot (S_t - k) - 1} \right\rfloor - k > 0$$

Choosing, without loss of generality, $t = 0$ and after some arithmetics, we see that:

$$S_0 > \left\lfloor \frac{k^2}{8 \, (1 - \delta)^2 \, p^2} + k + \frac{1}{2} \right\rfloor$$

As S_0 as a function of δ is monotonically increasing and tends to infinity as δ tends to 1, we can lower bound S_0 with $\delta = 0$, therefore:

$$S_0 > \left\lfloor \frac{k^2}{8 \cdot p^2} + k + \frac{1}{2} \right\rfloor \tag{5}$$

We would like this process to continue for t time steps, thus applying the same method for all time steps and using Definition 10, we get that:

$$\forall t \, Pr \, [S_t \geq S_0] = Q_t.$$

Notice that Theorem 5 produces two results - the first one is the minimal initial region's size S_0 which guarantees that the cleaning agents *will not* be able to successfully accomplish the cleaning process and second one is the corresponding probability Q_t in which this S_0 can be guaranteed. Also notice that in order to evaluate Q_t, one should use the appropriate S_0 and δ.

Interestingly, the results demonstrated in Fig. 13 of the impossibility result as presented in Theorem 5, as the number of cleaning agents increases the probability which we can guarantee the minimal initial region's size S_0 also increases - in other words, although as the number of agents increases, the corresponding minimal initial region's size also increases and the probability in which we can guarantee that the agents will not be able to successfully clean the region increases as well.

6.2 Markovian Model

Using the method presented in Sect. 5 bounding the probability for successful completion of the cleaning mission with probability P_t^{MIN}, performed by k simple cleaning agents we can also deduce the probability for failure in the cleaning mission, P_t^{MAX}.

Fig. 13 An illustration of
the impossibility result as
presented in Theorem 5. In **a**
we can see the minimal
initial region's size S_0 for
number of agents $k = [1..60]$
and spreading probability
$p = [0.1, 0.2, 0.3, 0.4]$. In **b**
we can see the corresponding
probability Q_t

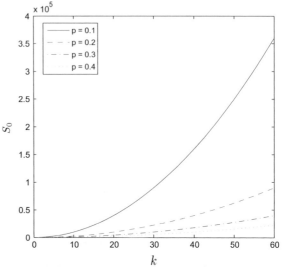

(a) The minimal initial region's size S_0 as a function
of the number of agents k for various probabilities.

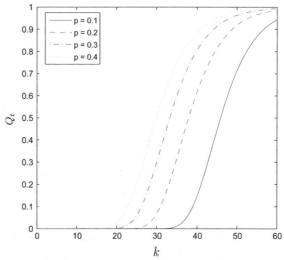

(b) The guaranteed probability Q_t as a function of the
number of agents k for various probabilities.

In the bound presented in Sect. 5 we assume that the contaminated region's size
is bounded from both directions. The lower bound S^{MIN} represents a successful
completion of the cooperative cleaning mission and the upper bound S^{MAX} represents
a total failure of this mission. One can ask for the probability for failing in the
cleaning mission for a given combination of the problem parameters - i.e. starting

Fig. 14 An illustration of the fail probability P_t^{MAX} using Theorem 4 as a function of the number of cleaning agents k and for various values of the spreading probability p. With $S^{MAX} = 5000$ and $S^{MIN} = O(k) > k$

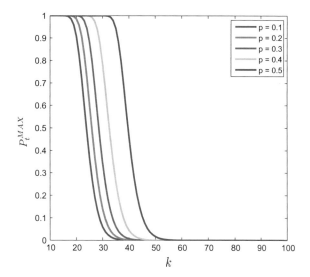

contaminated region's size, expending probability and number of cleaning agents. Using the power method in order to infer the stationary distribution of the probability to be in each one of the states between S^{MIN} to S^{MAX} as the time tends to infinity (or for a specific time step). Therefore we can compute the probability P_t^{MAX} to be in state S^{MAX} at time step t.

As we can notice in Fig. 14 for some of the smaller values of the number of cleaning agents k, the fail probability P_t^{MAX} is significant and tends to 1. This probability decreases as the number of cleaning agents increases.

7 Experimental Results

A computer simulation, implementing the **SWEEP** cleaning protocol as was suggested in [3], was constructed. Exhaustive simulations were carried out, examining the cleaning activity of the protocol for various combinations of parameters — namely, number of agents, spreading probability (or spreading time in the deterministic model) and geometric features of the contaminated region. All the results were averaged over at least 1000 deferent runnings in order to get a statistical significance. In the deterministic model we average the results over the all the possible starting position of the agents. Notice that, in order to minimize the running time, all running were stopped after some significant time and we consider these runnings as failure – i.e. these results are not included in the average calculation and not counted in the success percentage.

Fig. 15 Experimental
results for various number of
agents for spheric and
squared contaminated region
with starting size $S_0 = 500$
and with spreading
probability of $p = 0.02$
(notice that all the running
were stopped after 3000 time
steps). In **a** can see the
results compared to the
deterministic model results
(with $d = \frac{1}{p}$ and in **b** the
success percentage

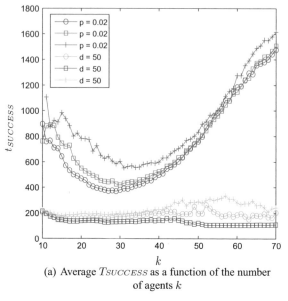

(a) Average $T_{SUCCESS}$ as a function of the number
of agents k

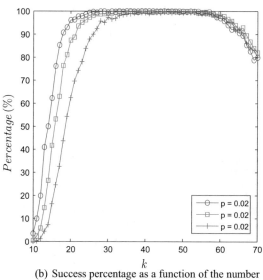

(b) Success percentage as a function of the number
of agents k

Some of the experimental results are presented in Fig. 15 comparing the probabilistic model and the deterministic model over three deferent shapes (circle, square
and cross) with range of number of agents. Notice the interesting phenomenon, where
adding more agents may cause to an increase in the cleaning time due to the delay
caused by the agents synchronization in the **SWEEP** protocol.

8 Conclusions

In this work we have studied the problem concerning a swarm of drone agents that is required to physically patrol a pre-defined (yet unknown) area, such that this area stochastically expands over time. This problem was thoroughly analyzed in previous works, assuming either a static search environment, or a deterministically expanding one (often referred to as the *Cooperative Cleaning* problem). We have presented several analytic bounds for the number of drones required in order to guarantee a successful "cleaning" of the region, as well as the time that would be required in order to complete this task. In addition, we have shown two impossibility results, allowing for the construction of problem scenarios that would be guaranteed to be insolvable by any patrolling algorithms that may be employed by the drones.

It is interesting to point out the relation of the *Markovian Model* presented in this work to the work presented in [6, 9–11], where a swarm of unmanned air vehicles (UAVs) is used for searching one or more "evading targets" which are moving in a pre-defined area while trying to avoid detection by the swarm. By arranging themselves into efficient geometric flight configurations, the UAVs optimize their integrated sensing capabilities, enabling the search of a maximal territory. In both works the areas which their shapes and sizes are unknown to the agents in advance and moreover those areas are bounded in a closed environment.

References

1. E.U. Acar, Y. Zhang, H. Choset, M. Schervish, A.G. Costa, R. Melamud, D.C. Lean, A. Gravelin, Path planning for robotic demining and development of a test platform, in *International Conference on Field and Service Robotics* (2001), pp. 161–168
2. L. Alonso, A.S. Goldstein, E.M. Reingold, 'lion and man': upper and lower bounds. Research Report RR-1700, INRIA. Projet PSYCHO ERGO (1992)
3. Y. Altshuler, A.M. Bruckstein, I.A. Wagner, Swarm robotics for a dynamic cleaning problem, in *IEEE Swarm Intelligence Symposium* (2005), pp. 209–216
4. Y. Altshuler, I.A. Wagner, A.M. Bruckstein, Shape factor's effect on a dynamic cleaners swarm, in *Third International Conference on Informatics in Control, Automation and Robotics (ICINCO), the Second International Workshop on Multi-Agent Robotic Systems (MARS)* (2006), pp. 13–21
5. Y. Altshuler, I.A. Wagner, A.M. Bruckstein, On swarm optimality in dynamic and symmetric environments. Economics. 7, 11 (2008)
6. Y. Altshuler, I.A. Wagner, A.M. Bruckstein, Collaborative exploration in grid domains, in *Sixth International Conference on Informatics in Control, Automation and Robotics (ICINCO)* (2009)
7. Y. Altshuler, I.A. Wagner, V. Yanovski, A.M. Bruckstein, Multi-agent cooperative cleaning of expanding domains. Int. J. Robot. Res. **30**, 1037–1071 (2010)
8. Y. Altshuler, V. Yanovski, D. Vainsencher, I.A. Wagner, A.M. Bruckstein, On minimal perimeter polyminoes, in *The 13th International Conference on Discrete Geometry for Computer Imagery (DGCI2006)* (2006), pp. 17–28
9. Y. Altshuler, V. Yanovski, I.A. Wagner, A.M. Bruckstein, The cooperative hunters - efficient cooperative search for smart targets using uav swarms, in *Second International Conference on*

Informatics in Control, Automation and Robotics (ICINCO), the First International Workshop on Multi-Agent Robotic Systems (MARS) (2005), pp. 165–170

10. Y. Altshuler, V. Yanovsky, A.M. Bruckstein, I.A. Wagner, Efficient cooperative search of smart targets using uav swarms. Robotica **26**, 551–557 (2008)

11. Y. Altshuler, V. Yanovsky, I. Wagner, A. Bruckstein, Swarm intelligencesearchers, cleaners and hunters, in *Swarm Intelligent Systems* (2006), pp. 93–132

12. Y. Altshuler, A. Bruckstein, On short cuts-or-fencing in rectangular strips (2010), arXiv:1011.5920

13. Y. Altshuler, A.M. Bruckstein, Static and expanding grid coverage with ant robots: complexity results. Theor. Comput. Sci. **412**(35), 4661–4674 (2011)

14. Y. Altshuler, S. Dolev, Y. Elovici, N. Aharony, Ttled random walks for collaborative monitoring, in *NetSciCom, Second International Workshop on Network Science for Communication Networks*, vol. 3 (San Diego, CA, USA, 2010)

15. Y. Altshuler, A. Pentland, S. Bekhor, Y. Shiftan, A. Bruckstein, Optimal dynamic coverage infrastructure for large-scale fleets of reconnaissance uavs (2016), arXiv:1611.05735

16. Y. Altshuler, R. Puzis, Y. Elovici, S. Bekhor, A.S. Pentland, On the rationality and optimality of transportation networks defense: a network centrality approach, in *Securing Transportation Systems*, pp. 35–63

17. R.C. Arkin, T. Balch, *Artificial Intelligence and Mobile Robots*, Cooperative Multi Agent Robotic Systems (MIT Press, Cambridge, 1998)

18. M.A. Batalin, G.S. Sukhatme, Spreading out: a local approach to multi-robot coverage, in *6th International Symposium on Distributed Autonomous Robotics Systems* (2002)

19. R. Bejar, B. Krishnamachari, C. Gomes, B. Selman, Distributed constraint satisfaction in a wireless sensor tracking system, in *Proceedings of the IJCAI-01 Workshop on Distributed Constraint Reasoning* (2001)

20. F. Berger, A. Gilbers, A. Grne, R. Klein, How many lions are needed to clear a grid? Algorithms **2**(3), 1069–1086 (2009)

21. G.D. Birkhoff, Proof of the Ergodic Theorem. Proc. Natl. Acad. Sci. USA **17**(12), 656–660 (1931)

22. R. Borie, C. Tovey, S. Koenig, Algorithms and complexity results for pursuit-evasion problems, in *Proceedings of the International Joint Conference on Artificial Intelligence (IJCAI)* (2009), pp.59–66

23. V. Braitenberg, *Vehicles* (MIT Press, Cambridge, 1984)

24. Z. Butler, A. Rizzi, R. Hollis, Distributed coverage of rectilinear environments, in *Proceedings of the Workshop on the Algorithmic Foundations of Robotics* (2001)

25. G. Chalkiadakis, E. Markakis, C. Boutilier, Coalition formation under uncertainty: bargaining equilibria and the bayesian core stability concept, in *AAMAS '07: Proceedings of the 6th International Joint Conference on Autonomous Agents and Multiagent Systems* (ACM, New York, NY, USA, 2007), pp. 1–8

26. S.A. DeLoach, M. Kumar, Multi-agent systems engineering: an overview and case study.*Intelligence Integration in Distributed Knowledge Management* (Idea Group Inc (IGI), 2008), pp. 207–224

27. G. Dudek, M. Jenkin, E. Milios, D. Wilkes, Robotic exploration as graph construction. IEEE Trans. Robot. Autom. **7**, 859–865 (1991)

28. A. Felner, Y. Shoshani, Y. Altshuler, A.M. Bruckstein, Multi-agent physical a* with large pheromones. J. Auton. Agents Multi-Agent Syst. **12**(1), 3–34 (2006)

29. J.O. Flynn, Lion and man: the general case. SIAM J. Control **12**, 581–597 (1974)

30. A.S. Goldstein, E.M. Reingold, The complexity of pursuit on a graph. Theor. Comput. Sci. **143**, 93–112 (1995)

31. R.P. Isaacs, *Differential Games: A Mathematical Theory with Applications to Warfare and Pursuit, Control and Optimization* (Wiley, New York, 1965). Reprinted by Dover Publications 1999

32. V. Isler, S. Kannan, S. Khanna, Randomized pursuit-evasion with local visibility. SIAM J. Discret. Math. **20**, 26–41 (2006)

33. W. Kerr, D. Spears, Robotic simulation of gases for a surveillance task, in *Intelligent Robots and Systems (IROS 2005)* (2005), pp. 2905–2910
34. S. Koenig, Y. Liu, Terrain coverage with ant robots: a simulation study, in *AGENTS'01* (2001)
35. S. Koenig, M. Likhachev, X. Sun, Speeding up moving-target search, in *AAMAS '07: Proceedings of the 6th International Joint Conference on Autonomous Agents and Multiagent Systems* (ACM, New York, NY, USA, 2007), pp. 1–8
36. S. Mastellone, D.M. Stipanovi, C.R. Graunke, K.A. Intlekofer, M.W. Spong, Formation control and collision avoidance for multi-agent non-holonomic systems: theory and experiments. Int. J. Robot. Res. **27**(1), 107–126 (2008)
37. T.W. Min, H.K. Yin, A decentralized approach for cooperative sweeping by multiple mobile robots, in *IEEE/RSJ International Conference on Intelligent Robots and Systems* (1998), pp. 380–385
38. K. Passino, M. Polycarpou, D. Jacques, M. Pachter, Y. Liu, Y. Yang, M. Flint, M. Baum, *Cooperative Control for Autonomous Air Vehicles, chapter Cooperative Control and Optimization* (Kluwer Academic, Boston, 2002)
39. M. Polycarpou, Y. Yang, K. Passino, A cooperative search framework for distributed agents, in *IEEE International Symposium on Intelligent, Control* (2010), pp. 1–6
40. R. Puzis, Y. Altshuler, Y. Elovici, S. Bekhor, Y. Shiftan, A.S. Pentland, Augmented betweenness centrality for environmentally-aware traffic monitoring in transportation networks
41. E. Regev, Y. Altshuler, A.M. Bruckstein, The cooperative cleaners problem in stochastic dynamic environments (2012), arXiv:1201.6322
42. I. Rekleitis, V. Lee-Shuey, A.P. Newz, H. Choset, Limited communication, multi-robot team based coverage, in *IEEE International Conference on Robotics and Automation* (2004)
43. J. Svennebring, S. Koenig, Building terrain-covering ant robots: a feasibility study. Auton. Robots **16**(3), 313–332 (2004)
44. B.T. Sebastian, Efficient exploration in reinforcement learning — technical report cmu-cs-92-102. Technical report, Carnegie Mellon University (1992)
45. D. Vainsencher, A.M. Bruckstein, On isoperimetrically optimal polyforms. Theor. Computut. Sci. **406**(1–2), 146–159 (2008)
46. R. Varga, *Matrix Iterative Analysis*, 1st edn. (Prentice Hall, Upper Saddle River, 1962)
47. I.A. Wagner, Y. Altshuler, V. Yanovski, A.M. Bruckstein, Cooperative cleaners: a study in ant robotics. Int. J. Robot. Res. (IJRR) **27**(1), 127–151 (2008)
48. I.A. Wagner, A.M. Bruckstein, From ants to a(ge)nts: a special issue on ant–robotics. Ann. Math. Artif. Intell. Special Issue on Ant Robot. **31**(1–4), 1–6 (2001)
49. I.A. Wagner, M. Lindenbaum, A.M. Bruckstein, Efficiently searching a graph by a smell-oriented vertex process. Ann. Math. Artif. Intell. **24**, 211–223 (1998)
50. *The Algebraic Eigenvalue Problem*, vol. 87 (Clarendon Press, Oxford, 1965)
51. X. Zheng, S. Koenig, Robot coverage of terrain with non-uniform traversability, in *In Proceedings of the IEEE International Conference on Intelligent Robots and Systems (IROS)* (2007), pp. 3757–3764

The Cooperative Hunters – Efficient and Scalable Drones Swarm for Multiple Targets Detection

1 Introduction

In the world of living creatures, "simple minded" animals such as ants or birds cooperate to achieve common goals with surprising performance. It seems that these animals are "programmed" to interact locally in such a way that the desired global behavior is likely to emerge even if some individuals of the colony die or fail to carry out their task for some other reasons. It is suggested to consider a similar approach to coordinate a swarm of collaborative drones without a central supervisor, by using only local interactions between the swarm's units. When this decentralized approach is used, much of the communication overhead (characteristic to centralized systems) is saved, the hardware of the vehicles can be fairly simple, and better modularity is achieved. A properly designed system should achieve reliability through redundancy, and potency through scale.

In addition, the scalability that is obtained by using a efficient decentralized design often results in a system whose performance can successfully compete with a centralized approach – who is significantly more expensive, less scalable, and prone to system-wide failures. Furthermore, in many cases (albeit not always), the performance of such systems is also much more easy to analyze.

In this work we suggest an efficient (and quasi-optimal) geometric approach for the design and analysis of a *Drones Swarm* – a collaborative system that is comprised of a multitude of collaborative Unmanned Air Vehicles (UAVs) – we call the *"Cooperative Hunters"*. This approach, presented in Sect. 5, uses optimized "flying patterns" for achieving an efficient hunt for any number of evading targets, maneuvering in a pre-defined rectangular area. The main advantages of the proposed design is its unlimited scale (manifested in the ability to use increasingly growing numbers of drones in the same basic design), and its ability to use any type of drones for this purpose – as long as any decrease in the velocity or sensing capabilities of the drones

---

This chapter is based on work previously published in parts in [7, 9, 10].

© Springer International Publishing AG 2018
Y. Altshuler et al., *Swarms and Network Intelligence in Search*,
Studies in Computational Intelligence 729, DOI 10.1007/978-3-319-63604-7_7

is appropriately compensated by an increase in the drones number (such that this increase requirement is analytically optimized).

Another important advantage of the proposed design is that it can successfully cope with targets that are faster than the drones, are smarter (in terms of computational resources, memory, etc.), and are equipped with longer-term sensors. These capabilities, as we show, fall short to the ability of the swarm to generate a synergic collaboration that trades resources in the individual level for numbers, and the ability to leverage them efficiently.

The efficiency of this new geometric flying configuration is analyzed, and compared both to techniques discussed in previous works and to the optimal solution for the problem.

2 Related Work

In recent years significant research efforts have been invested in design and simulation of multi-agent robotics and intelligent swarms systems — see e.g. [39, 63, 70] or [19, 38, 49] for biology inspired designs (behavior based control models, flocking and dispersing models and predator-prey approaches, respectively), [35, 58, 66, 71] for economics applications and [25, 41] for physics inspired approaches).

Tasks that have been of particular interest to researchers in recent years include synergetic mission planning [8], fault tolerance [54], swarm control [50], human design of mission plans [48], trends analysis [47, 53], marketing optimization [17, 18], role assignment [24], multi-robot path planning [72], traffic control [16, 57], formation generation [36], formation keeping [20], exploration and mapping [61], cleaning [69] and dynamic cleaning [2] and target tracking [62].

One of the most interesting challenges for a robotics swarm system is the design and analysis of a multi-robotics system for searching areas (whose dimensions and shape are either known or unknown) [21, 40, 45, 55, 60, 64] or see [1] for a survey of search and evasion strategies. Interesting works to mention in this scope are those of [43, 65], where a swarm of ant-like robots is used for repeatedly covering an unknown area, using a real time search method called *node counting*. By using this method, the robots are shown to be able to efficiently perform such a coverage mission, and analytic bound for the coverage time are discussed. Additional analysis of the effect various topological and geometric features of the search space have on the ability of a collaborative swarm to efficiently guarantee successful completion can be found in [3–5].

While in most works the targets of the search mission were assumed to be idle, recent works considered dynamic targets, meaning — targets which after being detected by the searching robots, respond by performing various evasive maneuvers intended to prevent their interception. In this context it is interesting to mention the roots of this field, dating back to World War II [52, 67]. The first planar search problem considered is the patrol of a corridor between parallel borders separated by width W. This problem was solved by in [44] in order to determine optimal patrol strategies for aircraft searching for ships in a channel.

A similar problem was presented in [68], where a system consists of a swarm of UAVs (unmanned air vehicles) was designed to search for one or more such "smart targets" (representing for example enemy units, or alternatively a lost friendly unit which should be a found and rescued). In this problem (presented in Sect. 3, the objective of the UAVs is to find the targets in the shortest time possible. While the swarm comprises relatively simple UAVs, lacking prior knowledge of the initial positions of the targets, the targets are equipped with strong sensors, capable of telling the locations of the UAVs from very long distances. The search strategy suggested in [68] defines *flying patterns* which the UAVs follow, designed for scanning the (rectangular) area in such a way that the targets cannot re-enter sub-areas which were already scanned by the swarm, without being detected (a summary of this solution is presented in Sect. 4).

The work of [51] discusses a similar scenario, where a single pursuing drone is required to engage evading targets in three different use-cases: the search for mobile targets through a linear corridor, the search for or bounding of a mobile target which starts inside of a circular region, and the prevention of a mobile target from entering a circular region.

Similar work, albeit in an underwater environment, is discussed in [26], whereas the work of [46] discusses the coordination of a team of autonomous sensor platforms searching for lost targets under uncertainty. A real-time receding horizon controller in continuous action space is developed based on a decentralized gradient-based optimization algorithm and by using the expected observation as an estimate of future rewards.

In [32] the author discusses a swarm of autonomous vehicles that patrols a water border. Optimization is reduced to determining the shape and parameters of the maneuvering trajectory and choosing the patrolling velocity.

A more technical problem concerning the way a swarm of drones can maneuver cooperatively is analyzed in [33] with is concerned with the problem of coordinating a team of mobile autonomous sensor agents performing a cooperative mission while explicitly avoiding inter-agent collisions in a team negotiation process. The work is a continuation of a previous work by the same authors [34].

Another work called *Exploration as Graph Construction*, provides a coverage of degree bounded graphs in $O(n^2)$ time, is described in [29]. Here a group of ant robots with a limited capability explores an unknown graph using special "markers". Another mechanism of using "pheromone markers" for the purpose of intra-swarm synchronization is discussed in [31], whereas a work that discusses a stochastic collaborative exploration of a virtual network for malicious cyber threats is discussed in [14].

Additional recent advancements in this field can be found in works such as [15, 23, 27, 28, 30, 56, 73], whereas works that analyzed the computational complexity of the problem can be found in [6, 12, 13, 22, 59].

3 The Cooperative Hunters Problem

As described above, the Cooperative Hunters problem discusses a swarm of UAVs utilized for searching and intercepting a set of evading targets. Following are more details concerning this model, as presented in [68]. Note that in the model described below several "real life" properties (such as sensors' errors or search regions which have curved topographies) were intentionally omitted, in order to provide a clear view of the geometric analysis.

3.1 The Targets

Definition 1 Let V_{target} denote the (known) maximum velocity of the targets.

Aside from knowing the targets' maximal speed, the UAVs possess no additional information regarding each target's actual speed, location or planned course. Each target is free to adjust its course and speed and is, in fact, assumed to be capable of intelligent evasion. Each target may spot the searching UAVs from a distance (far beyond the UAVs' detection range) and subsequently maneuver in an attempt to evade detection. This work focuses on the detection task only, leaving the actual tracking and handling of the targets once discovered out of its scope.

3.2 UAVs and Sensors

To carry out the search, a group of identically configured UAVs is given.

Definition 2 Let V_{UAV} denote the (constant) velocity of the UAVs.

We assume that each aircraft can detect targets located within its sensors' range.

Definition 3 Let D denote the detection diameter of the aircrafts.

Specifically, the aircraft will always detect a target that is placed within a radial distance of $\frac{D}{2}$ from it, and will never detect targets which are located beyond this range.[1]

The UAVs are equipped with Global Positioning System (GPS) receivers so that they can accurately ascertain the coordinates of their location at any time. Additionally, each UAV is aware of the geographic boundaries of the search region.

---

[1] As in reality the sensors have finite detection time, the value of D used in the model can be slightly smaller than the actual sensors' range, in order to generate positive overlap between a pair of UAVs moving in parallel (see more details in Sect. 5).

3.3   The Search Region

The search region is assumed to be known in advance, that is — the targets are known to be confined to a specific finite planar region. This assumption originates from the fact that a vehicle convoy might, for example, be surrounded by mountains, water and other features that restrict its location to a large (but limited) area. Similarly, a small boat patrolling crowded shipping lanes might have a fuel supply that restricts how far from the coastline it might proceed, or alternatively — a small yacht lost at a lake which should be found and rescued can be assumed to stay in water.

Simply put, each target's location is confined at all times to a rectangular region of width X, length Y and area $A = X \cdot Y$. Without loss of generality, the label X is assigned to the shorter side of the region (i.e. $X \leq Y$) and note that X is usually considerably larger than the sensor's detection diameter D. Apart from being confined within the rectangular boundaries of the search region, no other information concerning the targets' locations is available.

3.4   The Goal

The objective of this work is to develop an efficient search methodology for employing a UAV swarm in order to efficiently locate intelligent and evading targets within the search region. The search algorithm should attempt to maximize the probability of targets' detection and minimize the expected search time while minimizing the number of UAVs required. The algorithm should adapt in the face of UAV failures by reconfiguring the UAVs to optimally continue the mission with the surviving assets. The execution of the algorithm (including any adaptations necessitated by the loss of UAVs) should be accomplished with a minimum amount of control information to be passed between UAVs.

4   Previous Results

A search algorithm for the UAVs was presented in [68], designed to both limit the amount of information that must be exchanged between UAVs and to use simple algorithms to modify the search in the event of a UAV loss. This solution uses a small group of pre-defined swarm flying patterns. The flying pattern guarantee a successful completion of the mission, while their limited number allows an efficient and automatic reconfiguration of the UAVs array, in the case of a UAV malfunction. The initial number of UAVs in the pattern is decided (by ground commanders) prior to launch, and each UAV is given the pattern, its position in the pattern, and the dictionary of allowable alternative patterns.

Fig. 1 A line formation of
the UAVs, demonstrating a
static detection area of $N \cdot D$

Fig. 2 An illustration of the
algorithm proposed by
Wincent et al. [68]

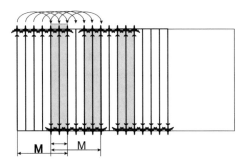

To achieve high searching efficiency, the UAVs form and maintain a configuration
of a straight line, as can be seen in Fig. 1. The lines formation then moves and scans
a portion of the rectangle (of size S) by moving *south*. Upon reaching the boundaries
of the search region, the formation travels a constant distance *eastwards*, and scans
another portion of the search region, by going *north*. Note that the two scanned areas
overlap in order to prevent targets from re-entering an area previously scanned. The
size of the overlap area is denoted as B. This process continues repeatedly, until the
eastern boundaries of the search area are met, at which time all targets are guaranteed
to be detected. A visual demonstration of the above is presented in Fig. 2.

The following result appears in [68] and demonstrates a lower bound over the
number of UAVs required to ensure a successful completion of the search mission:

Lemma 1 *Assuming we have N identically configured UAVs, moving at velocity
V_{UAV}, arranged in a line formation, such that their effective sweep width is $N \cdot D$,
searching for targets which are capable of moving at velocity V_{target}. Then, the
minimal number of UAVs required to ensure a forward-moving search with a detection
probability equals to one is given by:*

$$N_{min} = \min \left\{ \left\lfloor \frac{2X \cdot V_{target}}{V_{UAV} \cdot D} \right\rfloor + 1, \left\lceil \frac{Y}{D} \right\rceil \right\}$$

5 Improved Geometrical Flying Patterns

In this section we present improved flying patterns for the use by the UAVs, while
all the other aspects of the system as proposed in [68] can be maintained.

Let N, D, M, Y and X denote the number of UAVs, their sensors' detection
diameter, the line formation's scan width (note that $M = N \cdot D$), the length of the

Fig. 3 An illustration of the improved searching algorithm

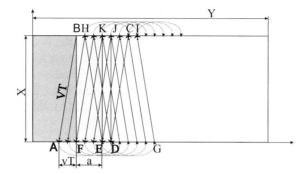

rectangular region and the width of the rectangular region, respectively. We now show an algorithm that can guarantee a successful detection of the targets with a much lower number of UAVs than required for the algorithm of [68].

5.1 Description of the Algorithm

Similarly to the algorithm of [68] we use a repeated scanning formation (north — south and vice versa), sweeping by a line formation of UAVs (as illustrated in Fig. 3).

Before each sweeping pass, the UAVs are located at either the northern or southern boundary of the search region, as a line formation. In Fig. 3 two sweeping passes are demonstrated. Without loss of generality let us assume that the first pass starts at time $t = 0$. Before the first and after the second pass the planes are at the northern boundary of the search region (a light-gray rectangle shows the addition to the "clean" area after the two passes).

The line formation first moves from the segment BC to AD. For the sake of simplicity, we assume the region west to the segment BF (appeared as dark gray in Fig. 3) to already be clean at time $t = 0$.

The points A and D towards which the UAVs formation heads are selected in such a way so that $\frac{|AB|}{V_{UAV}} = \frac{|AF|}{V_{target}}$ (thus preventing the targets from re-entering the already cleaned region west of the line AB).

Definition 4 Let T be defined as $T = \frac{|AB|}{V_{UAV}} = \frac{|AF|}{V_{target}}$.

Lemma 2 *When the formation reaches AD, the southern side of the rectangle at time $t = T$, the region to the west of line DJ is clean.*

Proof We partition the region to the west of line DJ to two parts and prove the Lemma for each of them separately:

1. No target can move to the west of line AB. Suppose a target crosses the line at some point $Z \in [AB]$. We assume that the target moved directly westwards, in order to reach a maximal distance from the UAVs (any other direction can be

treated as a westwards movement with a smaller V_{target}). Then, by the choice of A, if at time $t = 0$ the target was in $Q \in [BF]$, where Q is such that ZQ is parallel to axis X and the UAV at B, both the UAV and the target would reach Z simultaneously and the target will not cross AB undetected.

2. A target cannot move into $ABJD$. Indeed if it waits at line CD until the UAVs formation passes, and then heads westwards, by the choice of point A the target will reach line JD at the same time as the formation reaches the southern side of the area, and hence, will be detected.

When the formation reaches the segment AD, it shifts eastwards to occupy segment EG, in such a way that the area to the west of EK is guaranteed to be clean. Hence E is chosen such that $\frac{|AE|}{V_{UAV}} = \frac{|ED|}{V_{target}}$. This completes the description of one pass of the algorithm (as from EG the formation moves north to HI, then eastwards, etc.).

5.2 Analysis of the Algorithm

In this section we analyze the behavior of the algorithm for different values of V_{UAV}, V_{target} and X. First, we find the minimal ratio $\frac{V_{UAV}}{V_{target}}$, given X, for which a successful completion of the mission is guaranteed.

Theorem 1 If $\frac{V_{UAV}}{V_{target}} \geq \frac{X}{M} + 1$ then the UAVs formation is able to move eastwards while maintaining the area west to it clean.

Proof In order for the UAVs formation to be able to move eastwards while maintaining a clean area behind, each additional algorithm's pass must result in increasing the clean area to the west of the UAVs formation by some region of length $a > 0$. This can be maintained as long as the time it takes the formation to move from AD to EG is smaller than the time it takes a potential target to move from DJ to KE.

Hence, it must hold that $\frac{|AF|+a}{V_{UAV}} > \frac{|FD|-a}{V_{target}}$, when a is chosen such that $\frac{|BA|}{V_{UAV}} = \frac{|FA|}{V_{target}}$ (namely — the selected flying angle of the UAVs formation). That is, for the maximal possible propagation a we have:

$$\begin{cases} X^2 + (V_{target}T)^2 = (V_{UAV}T)^2 \\ \frac{V_{target}T+a}{V_{UAV}} = \frac{M-V_{target}T-a}{V_{target}} \end{cases} \tag{1}$$

which can also be written in the following way:

$$\begin{cases} T^2 = \frac{X^2}{V_{UAV}^2 - V_{target}^2} \\ a(V_{target} + V_{UAV}) = MV_{UAV} - V_{target}V_{UAV}T - V_{target}^2 T \end{cases} \tag{2}$$

As we are interested in assuring that $a > 0$, and as both $V_{UAV} > 0$ and $V_{target} > 0$, we know that also $a(V_{target} + V_{UAV}) > 0$, and as a result, using equations set (2) we know that

$$MV_{UAV} - V_{target}V_{UAV}T - V_{target}^2 T > 0$$

meaning that:

$$MV_{UAV} > (V_{target} + V_{UAV})V_{target}T$$

As both sides are positive, rewriting equations set (2) we get:

$$\begin{cases} T^2 &= \frac{X^2}{V_{UAV}^2 - V_{target}^2} \\ M^2 V_{UAV}^2 &> (V_{UAV} + V_{target})^2 V_{target}^2 T^2 \end{cases} \tag{3}$$

Assigning the value of T^2 into the second part of equations set (3) we get:

$$M^2 V_{UAV}^2 > (V_{UAV} + V_{target})^2 V_{target}^2 \frac{X^2}{V_{UAV}^2 - V_{target}^2}$$

which we shall rewrite as:

$$\frac{V_{UAV} - V_{target}}{V_{UAV} + V_{target}} \cdot \frac{V_{UAV}^2}{V_{target}^2} > \frac{X^2}{M^2} \tag{4}$$

Definition 5 Let us denote the velocities ratio as r, namely:

$$r = \frac{V_{UAV}}{V_{target}}$$

Therefore, Eq. (4) can now be written as:

$$\frac{r^2(r-1)}{r+1} > \frac{X^2}{M^2} \tag{5}$$

and, finally,

$$r^3 - r^2 - \frac{X^2}{M^2}r - \frac{X^2}{M^2} > 0 \tag{6}$$

A non-tight bound on r to satisfy Eq. (6) is:

$$r > \frac{X}{M} + 1 \tag{7}$$

Thus, given a sensor's detection diameter D, using the velocities ratio result of Expression (7), the minimal number of UAVs required to guarantee a successful completion of the mission can be obtained:

Theorem 2 *The minimal number of UAVs required for guaranteeing that the formation could move eastwards while maintaining the area west to it clean is:*

$$N_{min} = \min\left\{\left\lfloor \frac{X}{D} \cdot \frac{r+1}{r\sqrt{r^2-1}} \right\rfloor + 1, \left\lceil \frac{Y}{D} \right\rceil \right\}$$

Proof Let us rewrite Eq. (5) as follows:

$$r^2 \cdot (r-1) \cdot M^2 > X^2 \cdot (r+1)$$

which in turn can be written as:

$$r \cdot M > X \frac{r+1}{\sqrt{r^2-1}}$$

and the rest is consequentially implied.

Comparing Theorem 2 to the result of [68], presented in Lemma 1, yields the comparison between the performance of the two proposed algorithms:

Corollary 1 *Let $N_{min_{Vincent}}$ and $N_{min_{Hunter}}$ denote the minimal number of UAVs required to guarantee a successful completion of the mission using the algorithm presented in [68] and in Sect. 5.1, respectively. Then, excluding the trivial case of $N_{min_{Vincent}} = N_{min_{Hunter}} = \left\lceil \frac{Y}{D} \right\rceil$, we can see that:*

$$\frac{N_{min_{Vincent}}}{N_{min_{Hunter}}} \approx \frac{\frac{2X}{D \cdot r}}{\frac{X}{D} \cdot \frac{r+1}{r\sqrt{r^2-1}}} = 2\sqrt{\frac{r-1}{r+1}}$$

5.3 Lower Bound on the Number of UAVs — Optimality Proof

Once showing in the previous section that if $r > \frac{X}{M} + 1$, mission completion of a UAVs formation using the proposed algorithm can be guaranteed, we would now like to prove the following Theorem:

Theorem 3 *If $r < \frac{X}{M}$, the UAVs will not be able to complete their mission, regardless of the algorithm they employ.*

Proof For some given cooperative hunting algorithm, let $C(t)$ and $S(t)$ be defined as follows:

Definition 6 Let $C(t)$ denote the convex hull of the region guaranteed to be clean of targets at time t.

Definition 7 Let $S(t)$ denote the area of $C(t)$.

We shall now examine the behavior of $\frac{\partial S}{\partial t}$. Assuming that $r < \frac{X}{M}$, we show that if t is such that $S(t) = \frac{YX}{2}$ then $\frac{\partial S}{\partial t} < 0$, proving that the algorithm will not be able to complete its mission, once a certain targets-free region has been secured.

Definition 8 Let $P(t)$ denote the length of the circumference of $C(t)$ that is not part of the rectangle's boundary.

As the effective scanning region of the UAVs equals M, then during Δt time the UAVs can increase the area of the targets-free zone by at most $\Delta t M \cdot V_{UAV}$. As the targets can re-enter the targets-free zone from its boundaries (but not from the exterior of the bounding rectangle), they can decrease its area proportionally to its perimeter $P(t)$. Therefore, we can see that:

$$S(t + \Delta t) - S(t) < \Delta t M \cdot V_{UAV} - \Delta t P(t) \cdot V_{target}$$

By dividing both sides by Δt we get:

$$\frac{\partial S}{\partial t} \leq V_{target} \cdot (r \cdot M - P(t))$$

and by using the assumption that $\frac{X}{M} > r$ we can see that:

$$\frac{\partial S}{\partial t} < V_{target} \cdot (X - P(t))$$

We now need to show that $(S(t) = \frac{YX}{2}) \Rightarrow (X < P(t))$, which in turn is shown in Lemma 3 below.

Lemma 3 *For a convex region $C(t)$ contained entirely in a rectangle of dimensions X and Y (X being the short side), of area $S(t) = \frac{XY}{2}$, with sub-perimeter not touching the rectangle's boundaries $P(t)$, it holds that $P(t) \geq X$*

Proof We will divide the proof to several complementary parts, with respect to the number of sides of the search rectangle, which the convex region $C(t)$ touches.

- $C(t)$ does not touch any of the sides of the search region — as the shape of a given area which has the minimal perimeter is a sphere, it is clear that $P(t)$ is greater or equal to the perimeter of a sphere of an area $S(t)$, meaning that: $P(t) > \sqrt{2\pi XY}$. As $X \leq Y$ and since $\sqrt{2\pi} > 1$, we see that: $P(t) > X$.
- $C(t)$ touches a single side of the search region — by shifting the region towards one of the sides orthogonal to the side it currently touches (namely, and without loss of generality — eastwards if the region touches the northern or southern sides, and northwards for the eastern and western sides) a convex region with $P'(t) \leq P(t)$ is obtained. Using the next section discussing regions touching two sides of the search region, it can be seen that $P'(t) > X$, and therefore $P(t) > X$. See an example in Fig. 4.
- $C(t)$ touches two sides of the search region — in case the two sides are the north-south or east-west pairs, it is clear that $P(t) \geq 2X$. For any other couple of region's sides, let us consider Z, the triangle formed when dissecting the search region diagonally in such a way that the sides touching $C(t)$ are both part of Z.

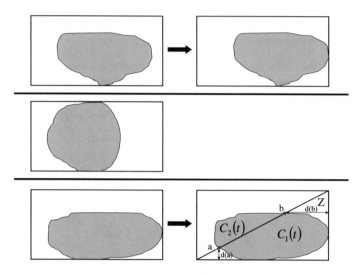

Fig. 4 An illustration of the second and third cases of Lemma 3

Since the area of Z equals $\frac{XY}{2}$, then since $S(t) = \frac{XY}{2}$ as well, either $Z = C(t)$ (in which case $P(t) = \sqrt{X^2 + Y^2} > X$) or all the following hold:

- $C(t) = C_1(t) \cup C_2(t)$
- $C_1(t) \cap C_2(t) = \emptyset$
- $Z \supset C_1(t)$
- $Z \cap C_2(t) = \emptyset$

Let us assume without loss of generality that $C(t)$ touches the western and southern sides of the search region. Let us denote by a and b the two points along the hypotenuse of Z, which are contained in $C(t)$, and which are the closest to the western and southern sides, respectively. Let us denote the distance between the western side to a by $d(a)$ and the distance between the southern side to b by $d(b)$. It is clear that $P(t) \geq d(a) + d(b) + P'(t)$ (where $P'(t)$ denotes the perimeter of $C_2(t)$, excluding the part of $C_2(t)$ which touches Z). As $P'(t) > \overline{ab}$ (the subsegment of the hypotenuses of Z which touches $C_2(t)$, which is located exactly between a and b, due to the convexity of $C(t)$), we can see that $P(t)$ has at least the length of some path beginning at the western side of the search region, proceeding to the hypotenuses of Z (at point a), going along the hypotenuses of Z until reaching the point b, and proceeding to the southern side of the search region.

We now show that the length of any path starting at the western side, going to the hypotenuses of Z, and returning to the southern side, is at least X:

It is obvious that the shortest path of such qualities will be in the form of two straight lines, one starting at some point c along the western side, going straight eastwards until reaching the hypotenuses of Z (denoted by e), and another going from this point straight southwards (denoted by f). Let $d(e)$ denote the length of e

and let $d(c)$ denote the distance between the western northern corner of the search region to point c. Note that $d(f) = X - d(c)$. The length of the path is therefore:

$$P = d(e) + X - d(c)$$

Note that since $\frac{d(c)}{d(e)} = \frac{X}{Y}$ the path's length can be written as:

$$P = d(e) + \left(X - \frac{X}{Y}d(e)\right) = d(e)\left(1 - \frac{X}{Y}\right) + X$$

It can easily be seen that for every value of $0 \le d(e) \le Y$ we see that $\frac{\partial P}{\partial d(e)} > 0$ (because $X \le Y$) and therefore the minimal value of P is produced for $d(e) = 0$, which is $P = X$. See an example in Fig. 4.

- $C(t)$ touches three sides of the search region — clearly, if the side not touched by $C(t)$ is the western or eastern, then $P(t) \ge X$ whereas if the side not touched by $C(t)$ is the northern or southern, then $P(t) \ge Y \ge X$.
- $C(t)$ touches all the four sides of the search region — Let k be the number of corners of the rectangle not contained in $C(t)$. Let a_i and b_i be the lengths of the portions of the sides adjacent to corner i that are not parts of $C(t)$ (a_i in the X coordinate and b_i in the Y coordinate). See an illustration of these notations in Fig. 5. Then the combined area of the triangles, formed by connecting these portions of the region's sides is:

$$\sum_{i=1}^{k} \frac{a_i b_i}{2}$$

As $C(t)$ is convex, by the definition of the triangles above, they contain all the area of the search region, which is not part of $C(t)$ (however, parts of the triangles may also be contained in $C(t)$). Therefore, the combined area of the triangles must be at least $\frac{XY}{2}$, and thus:

$$\sum_{i=1}^{k} \frac{a_i b_i}{2} \ge \frac{XY}{2}$$

Since $(a - b)^2 = a^2 + b^2 - 2ab$, and since $(a - b)^2 > 0$ for all $a, b \in \mathbb{R}$, we know that $a^2 + b^2 - 2ab > 0$ and therefore $a^2 + b^2 > 2ab$ and therefore $\frac{a^2 + b^2}{2} > ab$. Hence:

$$\sum_{i=1}^{k}\left(\frac{a_i^2 + b_i^2}{2}\right) \ge \sum_{i=1}^{k} a_i b_i \ge 2\sum_{i=1}^{k} \frac{a_i b_i}{2} \ge XY$$

and also:

$$\sum_{i=1}^{k}\left(a_i^2 + b_i^2\right) \ge 2XY$$

Fig. 5 An illustration of the
last case of Lemma 3. The
clean region in the center is
$C(t)$

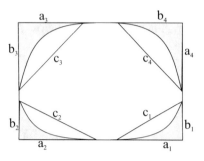

Denoting c_i the hypotenuses of the triangles, we get:

$$\sum_{i=1}^{k} c_i^2 \geq 2XY \geq 2X^2$$

As $\left(\sum_{i=1}^{k} c_i\right)^2 \geq \sum_{i=1}^{k} c_i^2$ we can write:

$$\left(\sum_{i=1}^{k} c_i\right)^2 \geq 2X^2$$

and therefore:

$$\sum_{i=1}^{k} c_i \geq \sqrt{2}X > X$$

As the sum of the hypotenuses of the triangles is a lower bound over $P(t)$ (since it is a sum of straight lines between pairs of points), we have shown that $P(t) > X$.

By producing this bound over the performance of a theoretical optimal algorithm, it is now interesting to compare this theoretical optimal algorithm to our scanning algorithm (namely, comparing the results of Theorems 3 and 2).

Corollary 2 *The efficiency of our algorithm in comparison to the theoretical optimal algorithm equals* $\sqrt{\frac{r-1}{r+1}}$

Proof Theorem 3 state $r = \frac{X}{M}$ as the theoretical lower bound for the velocities ratio r, required for enabling a successful scanning of a rectangular area (but not necessarily guaranteeing it), for any scanning algorithm. This can also be written as follows:

$$N = \frac{X}{r \cdot D}$$

Recalling the expression from Theorem 2, our proposed algorithm's efficiency is therefore approximately:

$$\frac{\frac{X}{r \cdot D}}{\frac{X}{D} \cdot \frac{r+1}{r\sqrt{r^2-1}}} = \frac{\sqrt{r^2-1}}{r+1} = \sqrt{\frac{r-1}{r+1}}$$

6 Conclusion

In this work we have presented an efficient and quasi-optimal flying pattern designed for a swarm of drones (or for that matter – any group of robots, capable of scanning a rectangular area freely) that guarantee the detection of any number of evading targets, as long as the number of drones is larger than an analytically given threshold, that is derived from the drones sensing capabilities as well as the ration between the velocities of the drones and the targets.

Let us examine once again the improvement of the scanning algorithm proposed in this work, in comparison to the results that are mentioned in [68]. The meaning of Corollary 1 can be best illustrated using a real life example. Assuming that an average UAV moves at a velocity of approximately 300 km/h, then for targets whose maximal velocity ranges between 6 km/h (e.g. a group of soldiers) and 70 km/h (an average armored vehicle, or a tank) the improvement factor of $2\sqrt{\frac{r-1}{r+1}}$ is translated to a saving of 49 and 37% from the number of UAVs required, respectively (i.e. ratios of 1.96 and 1.58).

Once showing that our algorithm is superior to this discussed in [68], we set out to investigate a possible optimal algorithm for the problem. The performance of a theoretical optimal algorithm is discussed in Theorem 3. This result is then compared to the performance of our algorithm, as appears in Corollary 2. Using the same "real life example" as discussed above, the efficiency of our algorithm would be 98% (for the group of soldiers moving at 6 km/h) and 79% (for the tanks moving at 70 km/h). Note that the optimality result holds for *any* algorithm. In other words, even drones that are equipped with a super computer which uses the A^* algorithm [37] for calculating their *optimal* flight patterns would not be able to achieve a performance improvement over our proposed designed that is beyond marginal.

Additional analysis of the geometric aspects of the optimization problem discussed in this work can be found in [11, 42].

References

1. S. Alpern, S. Gal, *The Theory of Search Games and Rendezvous* (Kluwer Academic Publishers, Dordrecht, 2003)
2. Y. Altshuler, A.M. Bruckstein, I.A. Wagner, Swarm robotics for a dynamic cleaning problem, in *IEEE Swarm Intelligence Symposium* (2005), pp. 209–216

3. Y. Altshuler, I.A. Wagner, A.M. Bruckstein, Shape factor's effect on a dynamic cleaners swarm, in *Third International Conference on Informatics in Control, Automation and Robotics (ICINCO), the Second International Workshop on Multi-Agent Robotic Systems (MARS)* (2006), pp. 13–21
4. Y. Altshuler, I.A. Wagner, A.M. Bruckstein, On swarm optimality in dynamic and symmetric environments. Economics **7**, 11 (2008)
5. Y. Altshuler, I.A. Wagner, A.M. Bruckstein, Collaborative exploration in grid domains, in *Sixth International Conference on Informatics in Control, Automation and Robotics (ICINCO)* (2009)
6. Y. Altshuler, I.A. Wagner, V. Yanovski, A.M. Bruckstein, Multi-agent cooperative cleaning of expanding domains. Int. J. Robot. Res. **30**, 1037–1071 (2010)
7. Y. Altshuler, V. Yanovski, I.A. Wagner, A.M. Bruckstein, The cooperative hunters - efficient cooperative search for smart targets using uav swarms, in *Second International Conference on Informatics in Control, Automation and Robotics (ICINCO), the First International Workshop on Multi-Agent Robotic Systems (MARS)* (2005), pp. 165–170
8. Y. Altshuler, V. Yanovsky, A.M. Bruckstein, I.A. Wagner, Efficient cooperative search of smart targets using uav swarms. Robotica **26**, 551–557 (2008)
9. Y. Altshuler, V. Yanovsky, I. Wagner, A. Bruckstein, Swarm intelligencesearchers, cleaners and hunters, in *Swarm Intelligent Systems* (2006), pp. 93–132
10. Y. Altshuler, *Multi Agents Robotics in Dynamic Environments*. Ph.D. thesis, Israeli Institute of Technology (2010)
11. Y. Altshuler, A. Bruckstein, On short cuts-or-fencing in rectangular strips (2010), arXiv:1011.5920
12. Y. Altshuler, A.M. Bruckstein, The complexity of grid coverage by swarm robotics, in *ANTS 2010* (LNCS, 2010), pp. 536–543
13. Y. Altshuler, A.M. Bruckstein, Static and expanding grid coverage with ant robots: complexity results. Theor. Comput. Sci. **412**(35), 4661–4674 (2011)
14. Y. Altshuler, S. Dolev, Y. Elovici, N. Aharony, Ttled random walks for collaborative monitoring, in *NetSciCom, Second International Workshop on Network Science for Communication Networks*, vol. 3 (San Diego, CA, USA, 2010)
15. Y. Altshuler, A. Pentland, S. Bekhor, Y. Shiftan, A. Bruckstein, Optimal dynamic coverage infrastructure for large-scale fleets of reconnaissance uavs (2016), arXiv:1611.05735
16. Y. Altshuler, R. Puzis, Y. Elovici, S. Bekhor, A.S. Pentland, On the rationality and optimality of transportation networks defense: a network centrality approach, in *Securing Transportation Systems* (2015), pp. 35–63
17. Y. Altshuler, E. Shmueli, G. Zyskind, O. Lederman, N. Oliver, A. Pentland, Campaign optimization through behavioral modeling and mobile network analysis. IEEE Trans. Comput. Soc. Syst. **1**(2), 121–134 (2014)
18. Y. Altshuler, E. Shmueli, G. Zyskind, O. Lederman, N. Oliver, A.S. Pentland, Campaign optimization through mobility network analysis, in *Geo-Intelligence and Visualization Through Big Data Trends* (2015), pp. 33–74
19. R.C. Arkin, Integrating behavioral, perceptual, and world knowledge in reactive navigation. Robot. Auton. Syst. **6**, 105–122 (1990)
20. T. Balch, R. Arkin, Behavior-based formation control for multi-robot teams. IEEE Trans. Robot. Autom. **14**(6), 926–939 (1998)
21. R. Bejar, B. Krishnamachari, C. Gomes, B. Selman, Distributed constraint satisfaction in a wireless sensor tracking system, in *Proceedings of the IJCAI-01 Workshop on Distributed Constraint Reasoning* (2001)
22. C. Bennett, On the nature and origin of complexity in discrete, homogeneous, locally-interacting systems. Found. Phys. **16**, 585–592 (1986). doi:10.1007/BF01886523
23. J. Cacace, A. Finzi, V. Lippiello, M. Furci, N. Mimmo, L. Marconi, A control architecture for multiple drones operated via multimodal interaction in search & rescue mission, in *2016 IEEE International Symposium on Safety, Security, and Rescue Robotics (SSRR)* (IEEE, 2016), pp. 233–239

24. C. Candea, H. Hu, L. Iocchi, D. Nardi, M. Piaggio, Coordinating in multi-agent robocup teams. Robot. Auton. Syst. **36**(2–3), 67–86 (2001)
25. D. Chevallier, S. Payandeh, On kinematic geometry of multi-agent manipulating system based on the contact force information, in *The Sixth International Conference on Intelligent Autonomous Systems (IAS-6)* (2000), pp. 188–195
26. H. Chung, E. Polak, J.O. Royset, S. Sastry, On the optimal detection of an underwater intruder in a channel using unmanned underwater vehicles. Naval Res. Logist. (NRL), **58**(8), 804–820 (2011)
27. M.G.C.A. Cimino, A. Lazzeri, G. Vaglini, Combining stigmergic and flocking behaviors to coordinate swarms of drones performing target search, in *2015 6th International Conference on Information, Intelligence, Systems and Applications (IISA)* (IEEE, 2015), pp. 1–6
28. M.G.C.A. Cimino, A. Lazzeri, G. Vaglini, Using differential evolution to improve pheromone-based coordination of swarms of drones for collaborative target detection, in *Proceedings of the 5th International Conference on Pattern Recognition Applications and Methods* (SCITEPRESS-Science and Technology Publications, Lda, 2016), pp. 605–610
29. G. Dudek, M. Jenkin, E. Milios, D. Wilkes, Robotic exploration as graph construction. IEEE Trans. Robot. Autom. **7**, 859–865 (1991)
30. M.A.A. El-Hadidy, Fuzzy optimal search plan for n-dimensional randomly moving target. Int. J. Comput. Methods **13**(06), 1650038 (2016)
31. A. Felner, Y. Shoshani, Y. Altshuler, A.M. Bruckstein, Multi-agent physical a* with large pheromones. J. Auton. Agents Multi-Agent Syst. **12**(1), 3–34 (2006)
32. A.A. Galyaev, E.P. Maslov, On the border patrolling problem. J. Comput. Syst. Sci. Int. **50**(5), 837 (2011)
33. S.K. Gan, R. Fitch, S. Sukkarieh, Real-time decentralized search with inter-agent collision avoidance, in *2012 IEEE International Conference on Robotics and Automation (ICRA)* (IEEE, 2012), pp. 504–510
34. S.K. Gan, S. Sukkarieh, Multi-uav target search using explicit decentralized gradient-based negotiation, in *2011 IEEE International Conference on Robotics and Automation (ICRA)* (IEEE, 2011), pp. 751–756
35. B.P. Gerkey, M.J. Mataric, Sold! market methods for multi-robot control. IEEE Trans. Robot. Autom. Spec. Issue Multi-robot Syst. **18**(5), 758–768 (2002)
36. N. Gordon, I.A. Wagner, A.M. Bruckstein, Discrete bee dance algorithms for pattern formation on a grid, in *IEEE International Conference on Intelligent Agent Technology (IAT03)* (2003), pp. 545–549
37. P.E. Hart, N.J. Nilsson, B. Raphael, A formal basis for the heuristic determination of minimum cost paths. IEEE Trans. Syst. Sci. Cybern. **4**(2), 100–107 (1968)
38. T. Haynes, S. Sen, Adaptation and learning in multi-agent systems, *Evolving Behavioral Strategies in Predators and Prey*, vol. 1042, Lecture Notes in Computer Science (Springer, Berlin, 1986), pp. 113–126
39. S. Hettiarachchi, W. Spears, Moving swarm formations through obstacle fields, in *International Conference on Artificial Intelligence* (2005)
40. W. Kerr, D. Spears, Robotic simulation of gases for a surveillance task, in *Intelligent Robots and Systems (IROS 2005)* (2005), pp. 2905–2910
41. S. Kirkpatrick, J.J. Schneider, How smart does an agent need to be? Int. J. Mod. Phys. C **16**, 139–155 (2005)
42. R. Klein, C. Levcopoulos, A. Lingas, Approximation algorithms for the geometric firefighter and budget fence problems, in *LATIN* (Springer, 2014), pp. 261–272
43. S. Koenig, Y. Liu, Terrain coverage with ant robots: a simulation study, in *AGENTS'01* (2001)
44. B. Koopman, *Search and Screening: General Principles with Historical Applications* (Pergamon Press, Oxford, 1980)
45. B.O. Koopman, The theory of search ii, target detection. Oper. Res. **4**(5), 503–531 (1956)
46. P. Lanillos, S.K. Gan, E. Besada-Portas, G. Pajares, S. Sukkarieh, Multi-uav target search using decentralized gradient-based negotiation with expected observation. Inf. Sci. **282**, 92–110 (2014)

47. Y.-Y. Liu, J.C. Nacher, T. Ochiai, M. Martino, Y. Altshuler, Prospect theory for online financial trading. PloS one **9**(10), e109458 (2014)
48. D. MacKenzie, R. Arkin, J. Cameron, Multiagent mission specification and execution. Auton. Robots **4**(1), 29–52 (1997)
49. M.J. Mataric, Designing emergent behaviors: From local interactions to collective intelligence, in *Proceedings of the Second International Conference on Simulation of Adaptive Behavior*, ed. by J. Meyer, H. Roitblat, S. Wilson (MIT Press, 1992), pp. 432–441
50. M.J. Mataric. *Interaction and Intelligent Behavior*. Ph.D. thesis, Massachusetts Institute of Technology (1994)
51. T.G. McGee, J.K. Hedrick, Guaranteed strategies to search for mobile evaders in the plane, in *Proceedings of the 2006 American Control Conference* (2006), pp. 14–16
52. P.M. Morse, G.E. Kimball, *Methods of Operations Research* (Wiley, MIT Press and New York, 1951)
53. W. Pan, Y. Altshuler, A. Pentland, Decoding social influence and the wisdom of the crowd in financial trading network, in *Privacy, Security, Risk and Trust (PASSAT), 2012 International Conference on and 2012 International Confernece on Social Computing (SocialCom)* (IEEE, 2012), pp. 203–209
54. L.E. Parker, Alliance: An architecture for fault-tolerant multi-robot cooperation. IEEE Trans. Robot. Autom. **14**(2), 220–240 (1998)
55. K. Passino, M. Polycarpou, D. Jacques, M. Pachter, Y. Liu, Y. Yang, M. Flint, M. Baum, *Cooperative Control for Autonomous Air Vehicles, chapter Cooperative Control and Optimization* (Kluwer Academic, Boston, 2002)
56. S. Perez-Carabaza, E. Besada-Portas, J.A. Lopez-Orozco, J.M. de la Cruz, A real world multi-uav evolutionary planner for minimum time target detection, in *Proceedings of the 2016 on Genetic and Evolutionary Computation Conference* (ACM, 2016), pp. 981–988
57. R. Puzis, Y. Altshuler, Y. Elovici, S. Bekhor, Y. Shiftan, A.S. Pentland, Augmented betweenness centrality for environmentally-aware traffic monitoring in transportation networks
58. G. Rabideau, T. Estlin, T. Chien, A. Barrett, A comparison of coordinated planning methods for cooperating rovers, in *Proceedings of the American Institute of Aeronautics and Astronautics (AIAA) Space Technology Conference* (1999)
59. E. Regev, Y. Altshuler, A.M. Bruckstein, The cooperative cleaners problem in stochastic dynamic environments (2012), arXiv:1201.6322
60. I. Rekleitis, V. Lee-Shuey, A. Peng Newz, H. Choset, Limited communication, multi-robot team based coverage, in *IEEE International Conference on Robotics and Automation* (2004)
61. I.M. Rekleitis, G. Dudek, E. Milios, Experiments in free-space triangulation using cooperative localization, in *IEEE/RSJ/GI International Conference on Intelligent Robots and Systems (IROS)* (2003)
62. B. Shucker, J.K. Bennett, Target tracking with distributed robotic macrosensors, in *Military Communications Conference (MILCOM 2005)*, vol. 4 (2005), pp. 2617–2623
63. L. Steels, Cooperation between distributed agents through self-organization, in *Decentralized A.I - Proc. first European Workshop on Modeling Autonomous Agents in Multi-Agents world* ed. by Y. DeMazeau, J.P. Muller (Elsevier, 1990), pp. 175–196
64. L.D. Stone, *Theory of Optimal Search* (Academic Press, New York, 1975)
65. J. Svennebring, S. Koenig, Building terrain-covering ant robots: a feasibility study. Auton. Robots **16**(3), 313–332 (2004)
66. S.M. Thayer, M.B. Dias, B.L. Digney, A. Stentz, B. Nabbe, M. Hebert, Distributed robotic mapping of extreme environments, in *Proceedings of SPIE, Mobile Robots XV and Telemanipulator and Telepresence Technologies VII* vol. 4195 (2000)
67. A. Thorndike, Summary of antisubmarine warfare operations in world war ii. Summary report, NDRC Summary Report (1946)
68. P. Vincent, I. Rubin, A framework and analysis for cooperative search using uav swarms, in *ACM Simposium on Applied Computing* (2004)
69. I.A. Wagner, Y. Altshuler, V. Yanovski, A.M. Bruckstein, Cooperative cleaners: a study in ant robotics. Int. J. Robot. Res. (IJRR) **27**(1), 127–151 (2008)

70. I.A. Wagner, A.M. Bruckstein, From ants to a(ge)nts: a special issue on ant–robotics. Ann. Math. Artif. Intell. Spec. Issue Ant Robot. **31**(1–4), 1–6 (2001)
71. M.P. Wellman, P.R. Wurman, Market-aware agents for a multiagent world. Robot. Auton. Syst. **24**, 115–125 (1998)
72. A. Yamashita, M. Fukuchi, J. Ota, T. Arai, H. Asama, Motion planning for cooperative transportation of a large object by multiple mobile robots in a 3d environment, in *In Proceedings of IEEE International Conference on Robotics and Automation* (2000), pp. 3144–3151
73. M. Zhang, J. Song, L. Huang, C. Zhang, Distributed cooperative search with collision avoidance for a team of unmanned aerial vehicles using gradient optimization. J. Aerosp. Eng. 04016064 (2016)

Optimal Dynamic Coverage Infrastructure for Large-Scale Fleets of Reconnaissance UAVs

1 Introduction

Last decade has seen a paradigmatic change in the operational processes of modern armies aerial forces, as designers and commanders have been gradually shifting their focus towards unmanned aerial vehicles (UAVs) [33, 33, 50]. Indeed, over 50% of the planes in the US today are unmanned, with this trend expected to further increase, and be adopted by additional western armies in the coming years [35, 80]. Israel, as a leader of UAV technology and military adoption [62, 70], is currently relying heavily on the use of UAVs in its military force, with such vehicles becoming an increasingly dominant element of its intelligence platforms. This has been the case in visual intelligent (VISINT), and recently also in tactical signal intelligence (SIGINT), which is traditionally in charge of 80% of the information gathered by the intelligence corps [2, 24].

Therefore, as the use of UAVs as an integral component on ongoing intelligence gathering in wartime, as well as during the "battle within the wars" increases, so grows the importance of the need to base this use on an efficient infrastructure. In other words, an innovative small scale dedicated UAV squadron designed for special missions may function perfectly with high redundancy and inefficient use of its resources, but a regular large-scale information gathering that is based on unmanned vehicles operating in swarms, cannot. Furthermore, the lack of an efficient infrastructure that assumes control of the low-level resource utilization tasks means that these tasks must ultimately be taken care of by the human operators (as is being done today) dramatically reducing the number of tasks these can engage, increasing the time it takes them to do so, as well as the overall cost of this process, and ultimately significantly limiting the vehicles operational potential.

As the complexity of the problem increases, so does the impact of optimizing capabilities on the overall resources required in order to guarantee a pre-defined level of performance. In other words, a successful use of large scale swarms of

This work was previously published in [20].

Y. Altshuler et al., *Swarms and Network Intelligence in Search*,
Studies in Computational Intelligence 729, DOI 10.1007/978-3-319-63604-7_8

UAVs as a combat and intelligence gathering tool necessitates the development of an efficient mechanism for optimization of their utilization, specifically in the design and maintenance of their patrolling routes.

This work proposes an efficient and robust analytic infrastructure for the deployment of collaborative drone swarms, focusing on its application for tactic intelligence gathering. Specifically, we present an analytic model for devising an optimal reconnaissance strategy for any given threat scenario, defined as: (a) The correlation function between the cost of a single drone and its detection performance; (b) The deployment method used (represented as a monotonically increasing function that models the coverage percentage as a function of the number of units); and (c) The estimated expected cost of an undetected threat.

In other words, for any deployment method of the drones swarm (three of which are discussed in this paper), and an answer to the question "how much detection performance do I get by using a more expensive units" the model generates an optimal strategy that is *threat-specific*, namely — economically optimizing the type and quantity of drones with respect to the cost of an undetected threat.

The proposed technique enables swarms of semi-autonomous vehicles to perform an efficient ongoing dynamic patrolling and scanning of the entire roads and transportation infrastructure in a pre-defined "search region", in a robust and near-optimal way, while guaranteeing detection of targets that are traveling in that region.

We demonstrate the applicability of our model using a comprehensive roads network of the Israeli roads and highways system, containing over 15,000 directed links. The rest of the paper is organized as follows: related work is presented in Sect. 2. The problem and the proposed patrolling model is presented in Sect. 3. Analysis is presented in Sects. 4 and 5, containing among others an analytic estimation of the required number of drones for a given threat scenario (Theorem 1), as well as an analytical estimation of the optimal types of drones to be used, if several drones of different costs and performances are available (Theorem 2). Section 6 presents a theoretical analysis of a case-study of the proposed monitoring method. Section 7 presents a second case study, using a real world transportation network dataset in order to discuss several deployment strategies for drones swarms to be used for different types of search regions. The data used for this analysis is presented and discussed in Sect. 8. Concluding remarks appear in Sect. 9. Readers interested in further expansion on the mathematical analysis of this problem and similar ones are invited to read [19, 21], which also partially overlap with this work.

2　Related Work

The need to efficiently monitor transportation networks has been the topic of many studies. For example, in [53], the optimal deployment of air quality monitoring units is discussed. An efficient placement of security monitors can be found in [66]. In land transportation, many works had focused on monitoring the transportation of hazardous materials. For example, infectious disease outbreaks pose a critical threat to public health and national security [29, 38]. Utilizing today's expanded trade and travel, infectious agents can be distributed easily within and across country borders as part of a biological terror attack, resulting in potentially significant loss of life, major economic crises, and political instability. Such threats stress even more the importance of a reliable and efficient transportation monitoring infrastructure. A survey of homeland security related threats and risks regarding transportation infrastructure can be found in [34]. In [67] the trade-off between accuracy and coverage, for given limited resources of sensor devices is discussed. Lam and Lo [60] proposed a heuristic approach to select locations for traffic volume count sensors in a roadway network. [98] proposed a sensor deployment framework to maximize such utilities. This framework has been extended to accommodate turning traffic information [37], existing installations and O-D information content [28], screen line problem [96], time-varying network flows [63] and unobserved link flow estimation [69].

In its military aspects and its relevant scientific literature, it is interesting to mention the roots of this field, dating back to World War II [65, 88]. The first planar search problem considered is the patrol of a corridor between parallel borders separated by a constant width. This problem aimed for determining optimal patrol strategies for aircraft searching for ships in a channel [57], with a more generic theory of optimal scanning that was later proposed in [68].

In recent years this problem had gained popularity with the emphasis on large scale swarms of drones [4, 5, 9, 12, 26, 54, 56, 68, 82, 89, 91], using a variety of methods ranging from the analysis of the geometrical properties of the search space [8, 10], the use of multi-agent robotics approaches [51, 84, 90] or [23, 49, 64] for biology inspired designs (behavior based control models, flocking and dispersing models and predator-prey approaches, respectively), network theory oriented architectures [20, 21, 71] or economics inspired designs [47, 76, 87, 94] or physics inspired approaches [36, 55] (see [3] for a survey of search and evasion strategies).

In general, most of the techniques used for the task of a distributed monitoring use some sort of cellular decomposition. For example, in [78] the area to be monitored is divided between "monitoring agents" based on their relative locations. In [32] a different decomposition method is being used, which is analytically shown to guarantee a complete coverage of the area. Another interesting work is presented in [1], discussing two methods for cooperative monitoring (one probabilistic and the other based on an exact cellular decomposition). Another work called *Exploration as Graph Construction*, provides a coverage of degree bounded graphs in $O(n^2)$ time, is described in [41]. Here a group of ant robots with a limited capability explores an unknown graph using special "markers". Another mechanism of using "pheromone markers" for the purpose of intra-swarm synchronization is discussed in [45], whereas a work that discusses a stochastic collaborative exploration of a virtual network for malicious cyber threats is discussed in [13].

While in most works the targets of the search mission were assumed to be idle, recent works considered dynamic targets, meaning — targets which after being detected by the searching robots, respond by performing various evasive maneuvers intended to prevent their interception. Some of these works can be seen in [6, 11, 14], whereas a variant that emphasizes the stochastic aspects of the problem can be found in [77].

3   Patrolling System Optimizing — Problem Definitions

This work analyzes the design of efficient monitoring systems through its economic perspective. Namely, when a multitude of threat scenarios are available, the main challenge is mitigating these scenarios in the cost-effective way. That is, resolving the real-time uncertainty regarding the exact kinds of threats through the optimization of its overall costs (both the direct costs of the system, as well as the indirect costs of undetected threats). It is therefore crucial to provide planners and operators with a monitoring system that can be reconfigured in real-time, having drones deployed/activated gradually as they are needed (due to new information regarding the unfolding situation, and taking into account other operational requirements or budget constraints).

Our proposed model does not assume any constraint on the deployment scheme itself, as the latter can be highly influenced by a large variety of considerations. Rather, it enables planners to model any given deployment scheme they desire, providing the optimal number and types of drones for each one. Our proposed model also offers operators high flexibility regarding the types of drones to be used — modeled using a trade off between the number of monitoring drones and their detection quality with respect to the number of potential targets each drone can monitor simultaneously (to be denoted as the drones' "*Sampling*" quality, a number ranging between 0 and 1). In this work we assume that a higher sampling quality implies a higher cost per drone. Therefore, the overall cost of the monitoring system can be modeled as:

$$Overall\ Cost = Cost\ per\ Unit\ \times\ Number\ of\ Units$$

whereas the overall monitoring performance of the system can be modeled as:

$$Overall\ Performance = Monitoring\ Coverage\ Percentage \times\ Sampling.$$

We note that the quality of monitoring units (i.e. their sampling rate) has in fact a *non linear* effect on the overall monitoring quality. For example, assuming the sampling rate of each monitor is $\frac{1}{2}$ then will be able to detect slightly more than half of the vehicles traveling in the network, because some of them might pass by more than one monitor and thus have a chance of being detected by either one. Specifically, such vehicles will have probability of $1 - (0.5)^2 = 0.75$ to be detected. Since drones' deployments is likely be done is much smaller numbers compared to the overall possible locations, it is unlikely that many potential targets would pass by two drones during a single trip. We can therefore use a linear approximation.

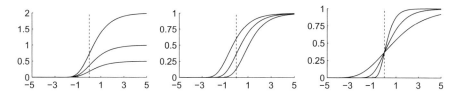

Fig. 1 An illustration of the *Gompertz function*. The charts represent the following functions (from *left* to *right*): $y = xe^{-e^t}$, $y = e^{-xe^t}$ and $y = e^{e^{-xt}}$, for $x = \frac{1}{2}$, $x = 1$ and $x = 2$

Definition 1 Let $C_{ATTACK}(\chi)$ denote the expected cost associated with the damage occurred by an attack χ, used to model an undetected target.

The estimation of the damage costs are beyond the scope of this paper.

Definition 2 Let $M(x) \in [0, 1]$ be a monotonically increasing function denoting the percentage of potential targets moving on the roads network, that is detected and analyzed using x monitoring units, using some given monitoring deployment scheme.

A monitoring system of n units would therefore be cost-effective as long as:

$$C_{ATTACK} \geq \frac{n}{M(n)} \cdot Cost\ per\ Unit \tag{1}$$

For modeling the function $M(x)$ we propose to use the well-known *Gompertz function* [48], whose general form is:

$$y(t) = ae^{be^{ct}}$$

The *Gompertz function* is widely used for modeling a great variety of processes, (due to the flexibly way it can be controlled using the parameters a, b and c), such as mobile phone uptake [79], population in a confined space [42], or growth of tumors [40] (see an illustration in Fig. 1). Its ability to model the progress of optimization process as a function of the available resources can also be seen in [15–18].

The function $M(x)$ can be extrapolated using simulations, as demonstrated in Fig. 13. Note that $M(x)$ is domain dependant and may significantly change for different networks.

In the next sections we show how this model can be used in order to calculate the optimal number and types of the monitoring drones.

4 The Optimal Number of Monitoring Units

Definition 3 The *Normalized Benefit* of a monitoring system of n units is defined as the expected monetary saving from preventing attacks from which the fixed cost of the monitoring units is subtracted:

$$\omega \triangleq C_{ATTACK} \cdot M(n) - n \cdot \text{Cost per Unit}$$

Note that this function can refer also to non-monetary aspects of a successful attack. It also resembles the "Net Present Value" calculation, but without explicitly accounting for discount rates.

Theorem 1 *For any values of* C_{ATTACK}, *a, b, c and* Cost per Unit, *the optimal number of monitoring units that would maximize the Normalized Benefit of a monitoring system is:*

$$n_1 = \frac{\ln\left(\frac{1}{a \cdot b \cdot c} \cdot \frac{\text{Cost per Unit}}{C_{ATTACK}}\right) - W\left(\frac{1}{a \cdot c} \cdot \frac{\text{Cost per Unit}}{C_{ATTACK}}\right)}{c}$$

$$n_2 = \frac{\ln\left(\frac{1}{a \cdot b \cdot c} \cdot \frac{\text{Cost per Unit}}{C_{ATTACK}}\right) - W_{-1}\left(\frac{1}{a \cdot c} \cdot \frac{\text{Cost per Unit}}{C_{ATTACK}}\right)}{c}$$

where $W(x)$ *is the* Lambert product log, *that can be calculated using the series:*

$$W(x) = \sum_{n=1}^{\infty} \frac{(-1)^{n-1} n^{n-2}}{(n-1)!} x^n$$

Proof The optimal value of the Normalized Benefit is received for the number of monitoring units that nullifies the derivative $\frac{\partial \omega}{\partial n}$:

$$\frac{\partial \omega}{\partial n} = C_{ATTACK} \cdot \frac{\partial M(n)}{\partial n} - \text{Cost per Unit}$$

Namely:

$$\frac{\partial M(n)}{\partial n} = \frac{\text{Cost per Unit}}{C_{ATTACK}} \tag{2}$$

(notice that we disregard interest rates, as both cost and damage can assumed to be subject to a similar change along time).

In that case, using Eq. (2) we obtain:

$$a \cdot b \cdot c \cdot e^{cn} \cdot e^{be^{cn}} = \frac{\text{Cost per Unit}}{C_{ATTACK}}$$

which in turn implies:

$$be^{cn} + cn - \ln \frac{\text{Cost per Unit}}{a \cdot b \cdot c \cdot C_{ATTACK}} = 0 \tag{3}$$

We note that $a > 0$ whereas $b, c < 0$. Analyzing Eq. (3) we can then see that in cases where:

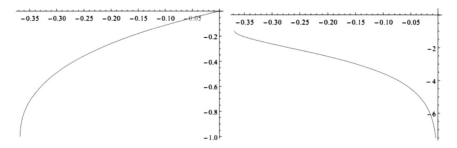

Fig. 2 The *left* and *right charts* depict the values of the *Lambert* functions $W(x)$ and $W_{-1}(x)$ in the segment $[-\frac{1}{e}, 0]$, respectively. The segment is implied by the constraint of Eq. (4)

$$\frac{\text{Cost per Unit}}{C_{ATTACK}} \leq -\frac{a \cdot c}{e} \qquad (4)$$

the optimal value of n would equal:

$$n_1 = \frac{\ln\left(\frac{1}{a \cdot b \cdot c} \cdot \frac{\text{Cost per Unit}}{C_{ATTACK}}\right) - W\left(\frac{1}{a \cdot c} \cdot \frac{\text{Cost per Unit}}{C_{ATTACK}}\right)}{c}$$

$$n_2 = \frac{\ln\left(\frac{1}{a \cdot b \cdot c} \cdot \frac{\text{Cost per Unit}}{C_{ATTACK}}\right) - W_{-1}\left(\frac{1}{a \cdot c} \cdot \frac{\text{Cost per Unit}}{C_{ATTACK}}\right)}{c} \qquad (5)$$

where $W(x)$ is the *Lambert product log*, and where $W_k(x)$ is its analytic continuation over the complex plane (the values of the functions $W(x)$ and $W_{-1}(x)$ in the segment implied by the constraint of Eq. (4) are illustrated in Fig. 2).

Returning to the Normalized Benefit of the system, we can now assign the values of the optimal number of monitoring units that appear in Eq. (5) into the definition of ω, as follows:

$$\omega_1 = C_{ATTACK} \cdot ae^{be^{cn_1}} - n_1 \cdot \text{Cost per Unit}$$

$$\omega_2 = C_{ATTACK} \cdot ae^{be^{cn_2}} - n_2 \cdot \text{Cost per Unit}$$

and using the properties of the W function, simplify it into the following form:

$$\omega_{max} = \max\{\omega_1, \omega_2\} \quad where: \qquad (6)$$

$$\omega_1 = a \cdot b \cdot C_{ATTACK} \cdot \gamma \cdot \left(W(b \cdot \gamma) + \frac{1}{W(b \cdot \gamma)} - \ln(\gamma)\right)$$

$$\omega_2 = a \cdot b \cdot C_{ATTACK} \cdot \gamma \cdot \left(W_{-1}(b \cdot \gamma) + \frac{1}{W_{-1}(b \cdot \gamma)} - \ln(\gamma)\right)$$

where the *Monitoring Benefit Factor* γ is defined as:

$$\gamma = \frac{1}{a \cdot b \cdot c} \cdot \frac{Cost\ per\ Unit}{C_{ATTACK}}$$

From Eq. (6) we can see that the Normalized Benefit of a drones monitoring swarm is a function of the *Monitoring Benefit Factor* γ, which takes into account both the parameters of the threat (through the overall potential damages and the costs of the monitoring units required to detect it) as well as the properties of the monitoring method (characterized by the values of a, b and c). Notice that the value of *Cost per Unit* does affect the values of a, b and c — as they are solely derived from the coverage efficiency of the network's traffic. However, we also note that the actual detection of the drones swarm is affected by their sampling quality, which in turn is monotonously affected by their cost.

5   Which Drones to Use? Optimizing the Drones' Cost

In this section we address the issue of finding the optimal type of monitoring drones that should be deployed. This is done by optimizing Eq. (6) with respect to the cost per single unit.

Definition 4 Let $Cost_{Base}$ denote the cost of the "most expensive" drone model. That is, the monitoring drone model with the highest level of detection available.

There is a wide range of monitoring drone models, where in most cases the cheapest ones are expected to have the poorest performance. We define the correlation between the sampling (or detection) quality and the cost of a single unit using a generic function, as follows:

$$Sampling = f_S\left(\frac{Cost\ Per\ Unit}{Cost_{Base}}\right) \tag{7}$$

Theorem 2 *For any values of* C_{ATTACK}, a, b, c, Cost per Unit, $Cost_{Base}$ *and for every function* $f_S : [0, 1] \to [0, 1]$, *the optimal cost for a monitoring unit that would maximize the Normalized Benefit of a monitoring system is a value that satisfies at least one of the following expressions:*

$$W(b \cdot \gamma) - \ln(\gamma) + \frac{1 - \frac{Cost\ Per\ Unit}{Cost_{Base}} \cdot \frac{\partial f_S}{\partial Cost\ Per\ Unit}\left[\frac{Cost\ Per\ Unit}{Cost_{Base}}\right] \cdot \left(f_S\left[\frac{Cost\ Per\ Unit}{Cost_{Base}}\right]\right)^{-1}}{W(b \cdot \gamma)} = 0$$

$$W_{-1}(b \cdot \gamma) - \ln(\gamma) + \frac{1 - \frac{Cost\ Per\ Unit}{Cost_{Base}} \cdot \frac{\partial f_S}{\partial Cost\ Per\ Unit}\left[\frac{Cost\ Per\ Unit}{Cost_{Base}}\right] \cdot \left(f_S\left[\frac{Cost\ Per\ Unit}{Cost_{Base}}\right]\right)^{-1}}{W_{-1}(b \cdot \gamma)} = 0$$

where:

$$\gamma = \frac{1}{a \cdot b \cdot c} \cdot \frac{\text{Cost per Unit}}{C_{ATTACK}} \cdot \left(f_S \left[\frac{\text{Cost Per Unit}}{\text{Cost}_{\text{Base}}} \right] \right)^{-1}$$

Proof We first revise Eq. (6) in order to take into account the different types of monitoring units as follows:

$$\omega_{max} = \max\{\omega_1, \omega_2\} \tag{8}$$

where ω_1 and ω_2 equal:

$$\omega_1 = a \cdot b \cdot C_{ATTACK} \cdot \text{Sampling} \cdot \gamma \cdot \left(W(b \cdot \gamma) + \frac{1}{W(b \cdot \gamma)} - \ln(\gamma) \right)$$

$$\omega_2 = a \cdot b \cdot C_{ATTACK} \cdot \text{Sampling} \cdot \gamma \cdot \left(W_{-1}(b \cdot \gamma) + \frac{1}{W_{-1}(b \cdot \gamma)} - \ln(\gamma) \right)$$

and where:

$$\gamma = \frac{1}{a \cdot b \cdot c} \cdot \frac{\text{Cost per Unit}}{C_{ATTACK} \cdot \text{Sampling}}.$$

We now proceed to the maximizing the financial merits of the monitoring system (namely, $\max\{\omega_1, \omega_2\}$). For this, we shall calculate the partial derivatives $\frac{\partial \omega_1}{\partial \text{Cost Per Unit}}$ and $\frac{\partial \omega_2}{\partial \text{Cost Per Unit}}$:

$$\frac{\partial \omega_1}{\partial \text{Cost Per Unit}} = \frac{1}{c} \cdot \left(W(b \cdot \gamma) - \ln(\gamma) + \frac{1 - \frac{\partial f_S}{\partial \text{Cost Per Unit}} \cdot \frac{\text{Cost Per Unit}}{\text{Cost}_{\text{Base}} \cdot f_S}}{W(b \cdot \gamma)} \right) \tag{9}$$

$$\frac{\partial \omega_2}{\partial \text{Cost Per Unit}} = \frac{1}{c} \cdot \left(W_{-1}(b \cdot \gamma) - \ln(\gamma) + \frac{1 - \frac{\partial f_S}{\partial \text{Cost Per Unit}} \cdot \frac{\text{Cost Per Unit}}{\text{Cost}_{\text{Base}} \cdot f_S}}{W_{-1}(b \cdot \gamma)} \right)$$

and the rest is implied.

Theorem 2 can now be used in order to calculate the *optimal aspired cost* of a single drone, for every correlation between its cost and its quality of detection, and for every deployment scheme and estimated threat's potential.

6 Case Study I – Theoretical Analysis

In this section we demonstrate how the proposed model can be used in order to produce the optimal number and types of monitoring drones, for a given threat scenario and drones deployment scheme, selected by the system's operators.

For the sake of simplicity we assume in this analysis that both sampling rates and Cost Per Unit to be continuous. This simplifies cases where the sampling rate of the monitoring units can be tuned resulting in lower resource utilization for lower

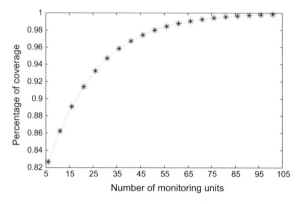

Fig. 3 An illustration of the *Gompertz* function $e^{-0.2 \cdot e^{-0.05 \cdot t}}$, used to simulate the increase in monitoring coverage as a function of the increase in number of drones, assumed in the case study of Sect. 6

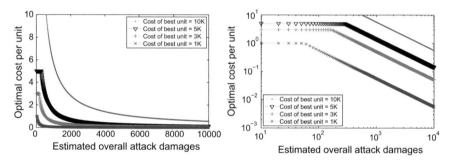

Fig. 4 An illustration of Eq. (11), denoting the optimal cost of a single drone, as a function of the estimations of the overall damages of a successful attack (from \$10K to \$10M). The optimal cost of a unit is shown for 4 possible values of $Cost_{Base}$, the cost of the best unit available. All monetary units are given in \$1K. Right chart uses a double log scale

sampling rates (e.g. $f_S(\frac{Cost\,Per\,Unit}{Cost_{Base}}) = \left(\frac{Cost\,Per\,Unit}{Cost_{Base}}\right)^2$). In addition, we assume that the deployment scheme chosen can be modeled with a *Gompertz* function of $a = 1$, $b = -0.2$ and $c = -0.05$ (see an illustration of this function in Fig. 3). In this case, nullifying the partial derivative of Eq. (9) would produce (Fig. 4):

$$\frac{\partial \omega_1}{\partial Cost\,Per\,Unit} = 0 \quad \longrightarrow \quad W(-0.2 \cdot \gamma) = \ln(\gamma) + \frac{1}{W(-0.2 \cdot \gamma)} \quad (10)$$

$$\frac{\partial \omega_2}{\partial Cost\,Per\,Unit} = 0 \quad \longrightarrow \quad W_{-1}(-0.2 \cdot \gamma) = \ln(\gamma) + \frac{1}{W_{-1}(-0.2 \cdot \gamma)}$$

subsequently implying:

$$\gamma_{opt} \approx 1.77356$$

(in this example, the optimal value of γ for ω_2 has a non-zero imaginary component).

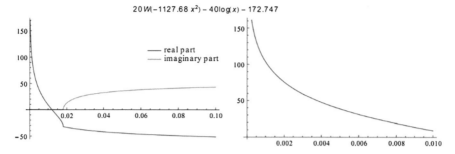

Fig. 5 An illustration of Eq. (12), showing the optimal number of drones as a function of the ratio between $Cost_{Base}$, the cost of the best drone model available, and the estimation regarding the overall damages of a successful attack (the ratio is denoted by x). The *left chart* contains both the real and imaginary parts of the solution. The *right chart* shows the segment between 0 and 0.01, where the imaginary part equals zero

Using this optimal value of γ we would now get:

$$\gamma_{opt} = \frac{1}{a \cdot b \cdot c} \cdot \frac{Cost\ per\ Unit}{C_{ATTACK} \cdot Sampling} = \frac{100 \cdot Cost_{Base}^2}{C_{ATTACK} \cdot Cost\ per\ Unit} = 1.77356$$

and from this we receive:

$$Cost\ per\ Unit_{opt} = \frac{100 \cdot Cost_{Base}^2}{1.77356 \cdot C_{ATTACK}} \approx 56.384 \cdot \frac{Cost_{Base}^2}{C_{ATTACK}} \qquad (11)$$

see an illustration of Eq. (11) in Fig. 5.

From Eq. (11) we can obtain for each kind of threat the optimal type of units that should be used, in order to maximize the Normalized Benefit of the system. In this example we also see a linear connection between the optimal cost of the drones and the overall estimations of the damages from a threat.

Assigning this back into Eq. (5), we can get the optimal number of units, for each potential threat (see an illustration in Fig. 5):

$$n_1 = 20 \cdot W\left(-1127.68 \cdot \frac{Cost_{Base}^2}{C_{ATTACK}^2}\right) - 172.747 - 40 \cdot \ln\left(\frac{Cost_{Base}}{C_{ATTACK}}\right)$$

$$n_1 = 20 \cdot W\left(-1, -1127.68 \cdot \frac{Cost_{Base}^2}{C_{ATTACK}^2}\right) - 172.747 - 40 \cdot \ln\left(\frac{Cost_{Base}}{C_{ATTACK}}\right)$$

$$\qquad (12)$$

Using Eqs. (11) and (12) we can now produce the optimal model of drones to be deployed, and in what numbers, for any given threat[1]: for example, if the best

_____

[1]Our model assumes continuous selection of drones' types. In reality of course there is a finite number of drones model. Therefore, after producing the optimal value for the cost of a single drone,

available drone model costs \$5,000 per unit then for threats of potential damages that are estimated as \$600,000 (which reflect a ratio of $\frac{1}{120}$ between the cost of best unit and the incident's damages) we should use approximately 25 units of the type that costs \$1,100 each (and deploy them according to the selected deployment scheme described above). However, for threats estimated at \$10,000,000 in total damages (i.e. a ratio of $\frac{1}{2000}$) an optimal monitoring system should be comprised of approximately 160 units, of the type that costs \$200 each.

Note that the previous example depends of course on the assumption of quadratic relation between cost and quality of the monitoring units, as well as on the assumption regarding the deployment scheme (i.e. the parameters of the *Gompertz Model*). For any change in any for these assumptions, new corresponding solutions to the optimal monitoring problem can be easily generated using the model.

7   Case Study II – Real World Transportation Network Monitoring

In the previous sections we have shown an analytic method for optimizing the number and types of monitoring units for any given deployment scheme (defined by the values of Gompertz function's parameters a, b and c that set the relation between the amount of monitoring resources and the increase in monitoring efficiency).

In this section we discuss several methods for deploying a given number of monitoring units. The goal in all three methods is the same — maximizing the probability that a target randomly moving in one of the network's edges, is detected.

The most trivial method for deploying n monitoring units is of course, the random deployment method. In other words, positioning the units in a randomly selected nodes in the network. As we will soon see, this method achieves surprisingly well performance in some cases. However, this method can be improved by taking into account certain features of the network, resulting in improved performance in many cases.

In the coming sections we will see several such improvements, all based on traffic-oriented improvement of the standard *Betweenness Centrality* method.

We then show that all methods shown in this section can be efficiently approximated using the *Gompertz* function $y = ae^{be^{ct}}$. Hence, any of those method can be represented by a set of values of a, b and c, as required by our proposed model. In addition, note that the last deployment method we present here achieves remarkable results (in terms of prediction of the real traffic flow, by the analysis of the network's topology). Therefore, this method can be assumed as a standard deployment scheme for transportation networks, even regardless of the scope of this work.

(Footnote 1 continued)
we would select the two available models closest in cost to this optimal value (namely, the more expensive cheaper one, and the cheapest more expensive one), assign their cost in the model, and select the one for which the merit function is higher.

The data used for the evaluation of this analysis was derived from a large number of location-reports, collected from cell phones in Israel, and is described and discussed in length in Sect. 8. Parts of the following analysis can also be found in [21, 75].

7.1 Betweenness Centrality Versus Traffic Flow

Betweenness Centrality (BC) stands for the ability of an individual node to control the communication flow in the networks and is defined as the total fraction of shortest paths between each pair of vertices that pass through a given vertex [22, 46]. In recent years Betweenness was extensively applied for the analysis of various complex networks [25, 86] including among others social networks [81, 93], computer communication networks [44, 97], and protein interaction networks [30]. Holme [52] have shown that Betweenness is highly correlated with congestion in particle hopping systems. Extensions of the original definition of BC are applicable for directed and weighted networks [31, 95] as well as for multilayer networks where the underlying infrastructure and the origin-destination overlay are explicitly defined [72].

Let $G = (V, E)$ be a directed transportation network where V is the set of junctions and E is the set of directed links as described in Sect. 8. Let $\sigma_{s,t}$ be the number of shortest paths between the origin vertex $s \in V$ and the destination vertex $t \in V$ (in some applications the shortest path constraint can be relieved to allow some deviations from the minimal distance between the two vertices). In the rest of this paper we will refer to the shortest or "almost" shortest paths between two vertices as *routes*. Let $\sigma_{s,t}(v)$ be the number of routes from s to t that pass through the vertex v. The Betweenness Centrality can hence be expressed by the following equation:

$$BC(v) = \sum_{s,t \in V} \frac{\sigma_{s,t}(v)}{\sigma_{s,t}}. \tag{13}$$

Note that in this definition we include the end vertices (s and t) in the computation of Betweenness since we assume that vehicles can be inspected also at their origin and at the point of their destination.

After computing the Betweenness Centrality for the given transportation network, we can easily see that the distribution of Betweenness Centrality follows a power law (Fig. 6). Long tail distributions such as the power law suggest that there is a non negligible probability for existence of vertices with very high Betweenness Centrality. This is in contrast to the exponential flow distribution depicted in Fig. 19. The different nature of these two distributions suggests that BC as defined above will overestimate the actual traffic flow through nodes especially for the most central vertices.

Next we would like to check the correlation between BC and traffic flow. Although the correlation is significant the square error is very low ($R^2 = 0.2021$) as shown in

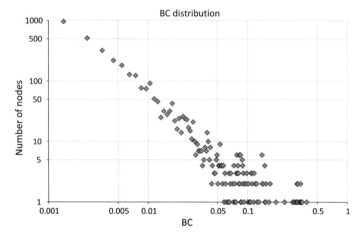

Fig. 6 Power law distribution of Betweenness Centrality

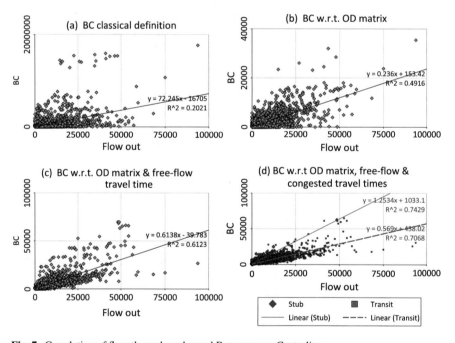

Fig. 7 Correlation of flow through nodes and Betweenness Centrality

Fig. 7a. Every point in this Figure represents a vertex with the x-axis corresponding to the measured traffic flow and y-axis corresponding to the computed BC.

We now discuss augmented variants of the Betweenness Centrality measure that significantly improve the correlation with the traffic flow.

7.2 Origin-Destination *Based Betweenness Centrality*

BC definition according to Eq. (13) BC assumes equal weights of routes between every pair of vertices in the network. In other words every vertex acts as an origin and as a destination of traffic. We would like to utilize the measured origin-destination (OD) flow matrix in order to prioritize network regions by their actual use. For this, we shall use the following altered definition for Betweenness, as suggested in [72]:

$$BC(v) = \sum_{s,t \in V} \frac{\sigma_{s,t}(v)}{\sigma_{s,t}} \cdot OD_{s,t} \qquad (14)$$

where OD is the actual measured origin-destination matrix. This method produces a better correlation ($R^2 = 0.4916$) between the theoretic (BC) and the measured traffic flow (see Fig. 7b).

7.3 *Improving Betweenness Centrality Using Travel Properties*

Shortest Routes based on Time to Travel In order to further improve our ability to estimate the predicted network flow using the network's topology, we note that both BC calculation methods (Eqs. (13) and (14) above) assume that routes are chosen according to shortest path strategy based on hop counting. In this section, we retain the shortest path assumption but use weighted links for calculating the Betweenness score. One option is to use the length of the road segments as their weights for the shortest path calculations. Shortest path algorithms (such as Dijkstra's or Bellman-Ford's) are able to consider only one distance weight on links when computing the shortest path to a destination. We shall therefore assume that the primary heuristic guiding people when they chose a route is the time required to reach their destination, and recompute the BC on the directed transportation, weighting links by their free-flow travel time.

Let $BC^{ft}(v)$ denote the Betweenness of a node v computed w.r.t. the free-flow travel time. Figure 7c shows significant improvements in the correlation between the measured traffic flow and the theoretical BC^{ft} values computed w.r.t the OD matrix and free-flow travel time link weights ($R^2 = 0.6123$). We can see that there are few nodes whose flow was significantly underestimated by the BC measure. Notice that there are also several nodes whose flow was actually overestimated. This can be explained by the fact that people do not travel strictly via shortest paths, but may have various deviations. In particular the deviations form shortest paths are affected by the day time and the day of week.

Peak-Hours Aware *Betweenness Centrality* It is a reasonable assumption that during peak hours travelers will choose to avoid the congested roads and choose their routes based on the congested travel time rather than on the free-flow travel times. Let

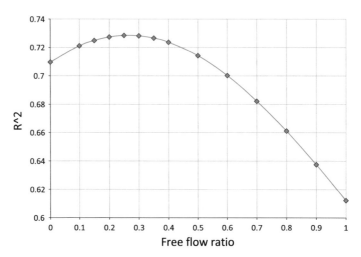

Fig. 8 Squared error (R^2) as the function of the free flow traffic fraction (α)

$BC^{ct}(v)$ denote the Betweenness of a node v computed w.r.t. the congested time. Computing Betweenness using only the congested travel time weights results in $R^2 = 0.7096$. Although peak hours are relatively small fraction of the day, most vehicles travel at these hours. This is the reason for higher correlation of BC^{ct} with the measured traffic flow.

We shall now combine both the Betweenness Centrality computed w.r.t. the free-flow travel time and the congested time by taking a weighted average, namely:

$$BC(v) = \alpha \cdot BC^{ft}(v) + (1 - \alpha) \cdot BC^{ct}(v)$$

where α denotes the relative fraction of vehicles traveling during the free-flow periods. The resulting centrality index can achieve higher correlation with the measured average traffic flow. The maximal correlation of $R^2 = 0.7285$ is obtained for $\alpha = 0.25$ as shown in the Fig. 8.

***Separating* Stubs Nodes *from* Transit Nodes** Nodes in the dataset were divided into two groups: *stub nodes* and *transit nodes*. A Stub node (i.e. centroid) is a node that is an origin or a destination of the traffic (as seen in the Origin-Destination matrix). These nodes account for approximately 10% of the network's nodes. All other nodes (namely, nodes that generate insignificant or no outgoing or incoming routes) are called Transit nodes, as they only forward traffic and do not generate or consume it.

Figure 7d presents the correlation that is received when the two groups of nodes are being processed separately. Specifically, the results show a $R^2 = 0.7068$ for the Transit nodes and a $R^2 = 0.7429$ for the Stub nodes.

Mobility Oriented *Betweenness Centrality* As each type of roads has a different functional class, we shall further improve our flow prediction by examining the Betweenness values achieved when calculating it for every group separately. The

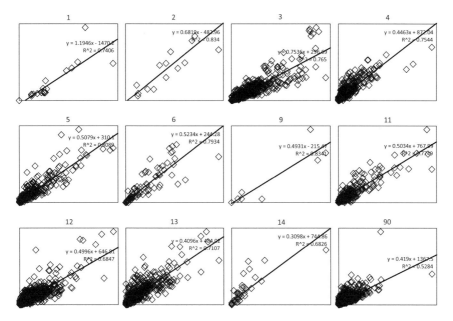

Fig. 9 Correlation of flow through nodes and Betweenness computed separately for different *types* of links

results of the correlation that is achieved using this method are presented in Fig. 9. We can clearly see that for the more important roads (namely, those with lower type number, representing a more infrastructural role in the transportation network) this technique yields R^2 values that are consistently above 0.74, reaching 0.83(!) for road of types 2 and 9 (note that roads of type 90 are fictive roads with infinite capacity that were artificially added in order to connect distinct regions in the network).

It should be noted that each node may have incoming roads of different types. Each plot corresponds to a set of nodes whose max incoming road type is as specified. In addition, the BC calculations were not made for each set of nodes separately — BC was computed for the complete network, while the correlations were computed separately for each type.

7.4 Optimizing the Locations of Surveillance and Monitoring Stations

In this section we use the Group variant of shortest path Betweenness Centrality (GBC) [43] as an estimate for the utility of collaborative monitoring. In other words, we are interested in verifying that given some mobile agent that we are interested in intercepting, we position the monitoring stations in a way that maximize the chance the agent would be captured, given the traffic patterns of the transportation network.

In this case, however, significant computational complexity issues arise, rendering the generation of an optimal solution impractical in real time by conventional tools that are based mostly on behavioral based modeling. Using Group Between Centrality we propose a way to generate efficient approximations of the optimal solution to this optimization problem.

GBC of a given group ($M \subseteq V$) of vertices accounts for all routes that pass through *at least one* member of the group. Let $\sigma_{s,t}$ and $\sigma_{s,t}(M)$ be the number of routes from s to t and the number of routes from s to t passing through at least one vertex in M respectively:

$$GBC(M) = \sum_{s,t \in V} \frac{\sigma_{s,t}}{\sigma_{s,t}(M)} \cdot OD_{s,t} \tag{15}$$

GBC can be efficiently computed using the algorithm presented in [73].

Assuming the routes are weighted by the origin destination flow in transportation networks, GBC will account for the net number of vehicles that are expected to pass by the monitors during an hour. This net number is different from the total number of vehicle passing by the monitors since the same vehicle can pass by several monitors during a single trip. For example, searching for a suspected escaping terrorist car, one would like to avoid stopping the same vehicle twice and increase the number of distinct vehicles that were inspected. It is therefore important to maximize the GBC value of the set of inspection stations given the number of stations deployed.

Several combinatorial optimization techniques can be used to find a group of nodes of given size that has the largest GBC. In the following discussion we refer to a greedy approximation algorithm for the monitors location optimization problem (Greedy) [39], a classical *Depth First Branch and Bound* (DFBnB) heuristic search algorithm [58], and recently proposed *Potential Search* [85].

The *Greedy* approximation algorithm chooses at every stage the node that has the maximal contribution to the GBC of the already chosen group. The approximation factor of the *Greedy* algorithm as reported in [39] is $e - \frac{1}{e}$.

Both the heuristic search algorithms *DFBnB* and the *Potential Search* provably find the group having the maximal GBC. The *Greedy* algorithm and *DFBnB* were previously compared in [74] in the context of monitoring optimization in computer communication networks. Given the fact that finding a group of a given size having the maximal GBC is a hard problem, the greedy algorithm is good enough for any practical purpose (the hardness of the problem can be proven by a straightforward reduction from the Minimal Vertex Cover problem that the problem of maximizing GBC is NP-Complete). Figure 10 presents the results of selecting one to 39 inspection locations using the greedy algorithm.

In certain cases (such as in various homeland security applications) deployment of monitoring systems are often done under tight timing conditions, as a result of new intelligence information. Therefore, any optimization method should provide close-to-real-time capabilities. In this context, it is interesting to note that both the *DFBnB* and the *Potential* algorithms are anytime search algorithms [99]. Their execution can be stopped at any point of time, yielding the best solution found so far. Therefore,

Fig. 10 The total net traffic flow that passes by monitors as a functions of the number of monitors. As expected the marginal value of additional monitors gradually decreases as more of them are added reaching potential traffic coverage of 30% when 39 monitoring stations are deployed

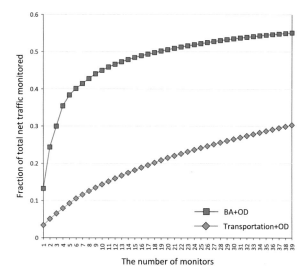

Fig. 11 The time (in seconds) that the search algorithms were executed as a function of the number of monitors. Note that due to the high complexity nature of the search algorithms, they present a phase transition around a tipping point (specific to each algorithm) in the number of monitors to be optimized

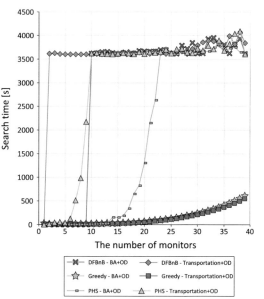

in the following experiments we limit the search time to one hour, simulating a quasi-real-time optimization constraint. Still, as can be seen in Fig. 11 the running time of then *Greedy* algorithm is by far lower than one hour, for the entire Israeli transportation system.

Fig. 12 The minimal quality of the solution (fraction of the upper bound) as a function of the number of monitors

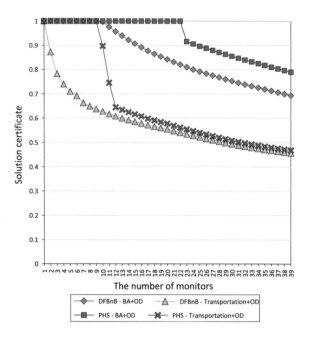

When *DFBnB* and *Potential Search* algorithms cannot complete the search process within the given time bounds they produce a close to optimal solution and an estimate of its optimality (i.e. certificate). The certificate is computed by dividing the best solution found so far by the upper bound on the optimal solution. The upper bound is computed using admissible heuristic functions and is maintained by the search algorithms for efficient pruning the search space. Figure 12 shows that *Potential Search* produces higher certificates for its solutions within the one hour time bound for all sizes of the monitors deployment.

7.5 *Modeling Various Deployment Schemes as a* **Gompertz** *Function*

Figure 13 demonstrates the performance of our monitoring method, by showing the percentage of traffic monitored as a function of the number of monitors, for several deployment schemes: (a) Group Betweenness, (b) Betweenness, and (c) Random deployment. The benefits of the proposed method can clearly be seen from this chart.

Remarkably, using GBC based deployment strategy it is possible to cover the vast majority of the traffic in the analyzed network (detecting any threat in probability

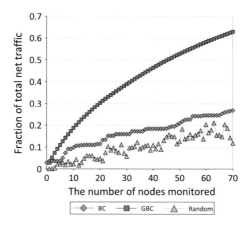

Fig. 13 The figure presents the results of deployment optimization performed on the Israeli transportation network with average travel times computed using state of the art traffic assignment model. Flows and the utility of the deployment were estimated using *Betweenness Centrality* and *Group Betweenness Centrality* models, and compared also to the random deployment model. Whereas the *BC* algorithm had chosen the locations for monitoring units according to the most central intersection based on their BC values, the *GBC* deployment was a greedy algorithm that tried to maximize the net-number of vehicles passing by the monitors. The benefits of the *GBC* strategy is clearly shown, as well as the ability to extrapolate this correlation between number of monitoring units and monitored traffic percentage, in order to find the minimal number of monitoring units required in order to guarantee certain levels of coverage

of approximately 0.7) using only 70 monitors. This result is most probably due to the relatively small number of origins and destinations in the analyzed network. 680 origins/destinations account for a little bit more than 10% of the network nodes.

BC based strategy produces relatively high quality deployments for small number of monitors (less than five). However, when 10 or more monitors need to be located random deployment is on average as effective as choosing the most central intersections. Moreover, for large numbers of monitors (more than 70–80) random deployment, although the simplest strategy, achieves coverage results that are very similar to choosing the most central intersections. This result may seem surprising but in fact it is absolutely reasonable. Central intersections tend to lay on the arterial roads and usually are quite close to each other. This results in reduced marginal utility of each additional junction joining the deployment.

We demonstrate Eq. (3) using the results presented in Fig. 13. For this, we need to calculate the regression of the simulated measurements presented in Fig. 13 to the *Gompertz* function. This is presented in Fig. 14. The regression yields the following results:

$$\textbf{Random deployment}: M(n) = 0.24e^{-2.73e^{-0.03n}} \tag{16}$$

$$\textbf{BC deployment}: M(n) = 0.51e^{-2.14e^{-0.02n}}$$

$$\textbf{GBC deployment}: M(n) = 0.89e^{-2e^{-0.04n}}$$

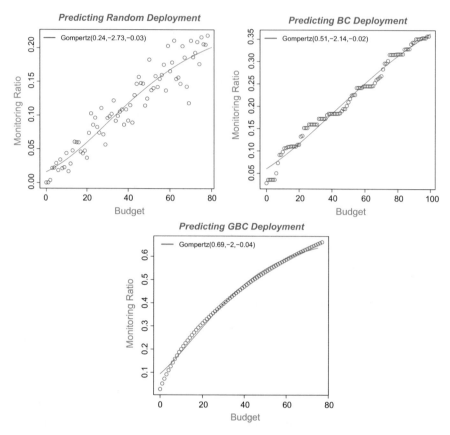

Fig. 14 *Gompertz* regression of the measurements presented in Fig. 13 — Random deployment, BC deployment and GBC deployment

At this point, we can assign the values of a,b and c in Eq. (3) and get the optimal number of monitoring units, for any ratio between the cost of a single monitoring unit and the expected cost of a successful attack. This is presented in Fig. 15, where the benefit of the GBC deployment scheme can clearly be seen, as it enables the use of much more expensive monitoring units (which make little sense in using, when the cost of using them is greater than the potential damage of a missed detection — due to low monitoring efficiency).

8   Transportation Network Dataset

In this section we evaluate our proposed model using a real-world transportation dataset, containing data regarding the Israeli roads structure, as well as the traffic through them (simulating the potential targets that need to be detected or monitored).

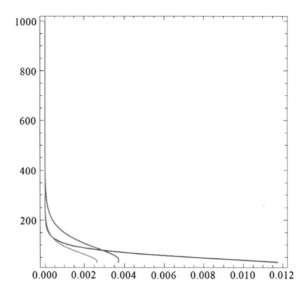

Fig. 15 The optimal number of monitoring units as a function of the ratio between the cost of a single monitoring unit, and the cost of a successful attack, for the *Gompertz* regressions presented in Fig. 14 for random deployment (*green*), BC deployment (*purple*), and GBC deployment (*blue*). Note how both the random deployment and the BC deployment schemes can be used only for very low-cost monitoring units, whereas the GBC deployment scheme enables the use of much more expensive monitoring units (due to its increased efficiency in guaranteeing high monitoring coverage)

This section presents the dataset and its various aspects, and discusses in great length its network properties – from which the optimal deployment of the drones swarm can be derived.

The widespread use of cellular phones in Israel enables the collection of accurate transportation data. Given the small size of the country, all cellular companies provide national wide coverage. As shown in [27], the penetration of cellular phones to the Israeli market is very high, even to lower income households, and specially among individuals in the ages of 10 to 70 (the main focus of travel behavior studies). Such penetration enables a comprehensive study of travel behavior that is based on the mobility patterns of randomly selected mobile phones in the Israeli transportation system. This data was shown in [27, 83] to provide a high quality coverage of the network, tracking 94% of the trips (defined as at least 2 km in urban areas, and at least 10 km in rural areas). The resulting data contained a wealth of traffic properties for a network of over 6,000 nodes, and 15,000 directed links. In addition, the network was accompanied with an Origin Destination (OD) matrix, specifying start and end points of trips.

The network was created for the National Israeli Transportation Planning Model. In urban areas the network contains arterial streets that connect the interurban roads. For each link of the network, there is information about the length (km), hierarchical type, free-flow travel time (min), capacity (vehicles per hour), toll (min), hourly flow

Table 1 Structural properties (Israeli transportation network)

Nodes	6716
Edges (undirected representation)	8374
Edges (directed representation)	15823
Number of structural equivalence classes	6655
Largest equivalence class	3
Number of bi-connected components (BCC)	931
Avg BCC size	8.2
Largest BCC	5778

(vehicles per hour), and congested travel time (min). The hourly flows and congested travel times were obtained from a traffic assignment model that loads the OD matrix on the network links.

8.1 Network Structure

Based on the dataset described above we have created a network structure, assigning running indices from 1 to 6716 to the nodes (junctions). We have examined the directed variant of the network where each road segment between two junctions was represented as either one or two directed links between the respective nodes.

In order to get a basic understanding of the network we first extracted and studied several of its structural properties (see Table 1). We have partitioned the network into structural equivalence classes of the nodes and bi-connected components and computed the Betweenness Centrality indices of the nodes (Betweenness Centrality is an important network feature that measures the portion of shortest paths between all the pairs of nodes in the network, that pass through a particular node. See more details in [46, 59, 61]). Structurally equivalent vertices have exactly the same neighbors and the set of these vertices is called a structural equivalence class. As can be seen from Table 1 the number of structural equivalence classes is roughly the number of vertices in the network and the size of the largest class is three. This means that there are no "star-like" structures in the network and alternative paths between any two vertices are either longer than two hops or have other links emanating from the intermediate vertices. On the other hand the number of biconnected components in the network is low compared to the number of nodes, meaning that there are significant regions of the network that can be cut out by merely disconnecting a single node (Fig. 16).

8.2 Congestions

In this paper we define the impact of congestion as the difference between the time to travel through a congested link and the free-flow time to travel. Congestion of a

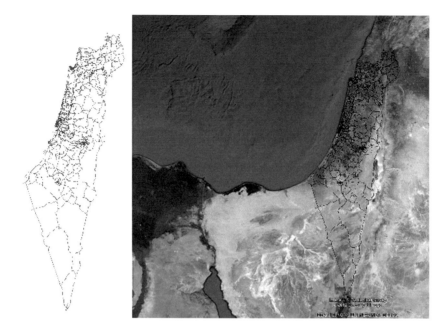

Fig. 16 A map of the Israeli transportation network that was used for this paper

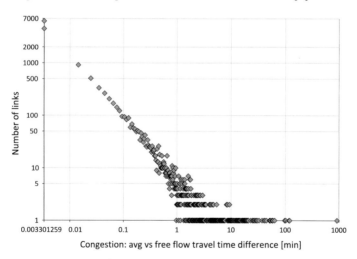

Fig. 17 Power law distribution of congestion

junction can be either inbound or outbound. Inbound congestion is the sum of all congestions on inbound links of some junction. Figure 17 presents the distribution of congestion on network nodes (junctions). Power law nature of this distribution means that vast majority of nodes are not congested but there are a few nodes whose congestion can be arbitrarily large. Based on the *Wardrop's User Equilibrium* [92] this also implies a low number of yet significant deviations between the routes chosen by travelers during free-flow and during congestions. In Sect. 7.3 we use this fact to merge between two routing strategies.

8.3 Flow

The analyzed dataset contains traffic flow through links provided as the number of vehicles per hour. We compute the total inbound flow through a node by summing flows on all of its inbound links, where outbound flow is computed symmetrically. Unless a specific junction is a source or a destination of traffic we expect the inbound flow to be equal to the outbound flow. Figure 18 demonstrates the correlation between inbound and outbound flow. We see that vast majority of the nodes are located on the main diagonal, however, there are some deviations, caused by the fact that the data represents average measurements that were carried out along a substantial period of time.

Figure 19 presents the distribution of inbound flow on network nodes. This distribution is exponential, meaning that a vast majority of nodes have little flow through them. However, in contrast to network congestion, there are no "unbounded fluctuations", i.e. the flow through the most "busy" junctions is not as high as can be expected from the power law distribution of betweenness and congestions (Figs. 17 and 6). In fact, congestions significantly limit the flow through the busiest junctions, which subsequently is the reason we do not see the long tail in flow distribution.

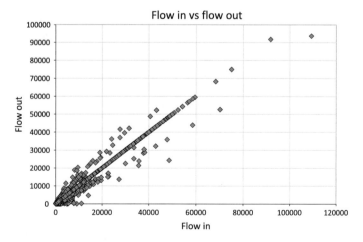

Fig. 18 Incoming vs. outgoing flow for each node

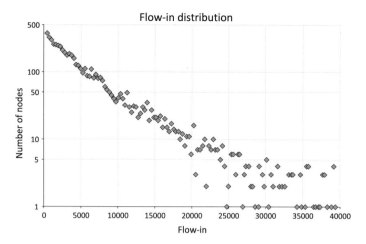

Fig. 19 Exponential distribution of traffic flow through nodes

9 Conclusions

In this paper we have discussed the problem of optimizing the type, number and locations of surveillance drones trying to detect maneuvering targets that move on top of a pre-defined transportation network. We have presented a model for analytically generating an optimal monitoring strategy, flexible enough to support various models of drones (of different costs and performance specifications), as well as any deployment scheme used. For each threat, the model produces an optimal strategy, based on the estimation of the threat's overall potential damages.

We have validated our model using a variety of deployment schemes, among which was an extremely efficient scheme we have developed, which is a transportation oriented variant of the Network Betweenness measure. We have then evaluated the model, as well as this deployment scheme, using a comprehensive dataset that covers the Israeli transportation network, and demonstrated how the optimal locations of any (reasonable) amount of drones can be approximated in high accuracy. The method proposed in this paper can now be used in order to generate highly efficient dynamic monitoring strategies, for a multitude of real-time threats, of different operational parameters.

A more theoretical approach to this problem that studies the complexity of all possible strategies for a collaborative monitoring of a given area at which an unknown number of targets dynamically maneuver can be found in [5]. An additional similar variant to this problem is the search for pollutant emitting vehicles, where the merit function is derived from environments considerations [75]. It is interesting to mentioned that in those variants as well, the topological properties of the network along which the "targets" can move significantly influence the ability of drones to track them, as was pointed out in [8, 10].

In future works we plan to conduct a thorough study of the way this method can be used for resource allocation scenarios involving several types threats, using heterogenous deployment strategies that would utilize drones of different characteristics simultaneously. In addition, we intend to conduct a further analysis of the resilience of the method, under uncertainty (regarding the topology, potential threats, and noises in the deployment scheme).

Finally, it is interesting to note that the problem of finding optimal (and optionally dynamic) monitoring strategy is related to other kinds of monitoring problems, such as "reverse-monitoring" – maximizing the exposure of a certain signal, or campaign to a population [19], or monitoring for evading land targets by a flock of Unmanned Air Vehicles (UAV). In this problem, however, the fact that the paths of the UAVs is unconstrained (as they are flying in the air) makes the calculation of a near-optimal monitoring strategy fairly easy [7, 11].

References

1. E.U. Acar, Y. Zhang, H. Choset, M. Schervish, A.G. Costa, R. Melamud, D.C. Lean, A. Gravelin, Path planning for robotic demining and development of a test platform, in *International Conference on Field and Service Robotics* (2001), pp. 161–168
2. M.M. Aid, All glory is fleeting: sigint and the fight against international terrorism. Intell. Natl. Secur. **18**(4), 72–120 (2003)
3. S. Alpern, S. Gal, *The Theory of Search Games and Rendezvous* (Kluwer Academic Publishers, Boston, 2003)
4. Y. Altshuler, A.M. Bruckstein, The complexity of grid coverage by swarm robotics, in *ANTS 2010* (LNCS, 2010), pp. 536–543
5. Y. Altshuler, A.M. Bruckstein, Static and expanding grid coverage with ant robots: complexity results. Theoret. Comput. Sci. **412**(35), 4661–4674 (2011)
6. Y. Altshuler, A.M. Bruckstein, I.A. Wagner, Swarm robotics for a dynamic cleaning problem, in *IEEE Swarm Intelligence Symposium* (2005), pp. 209–216
7. Y. Altshuler, V. Yanovski, I.A. Wagner, A.M. Bruckstein, The cooperative hunters - efficient cooperative search for smart targets using uav swarms, in *Second International Conference on Informatics in Control, Automation and Robotics (ICINCO), the First International Workshop on Multi-Agent Robotic Systems (MARS)* (2005), pp. 165–170
8. Y. Altshuler, I.A. Wagner, A.M. Bruckstein, Shape factor's effect on a dynamic cleaners swarm, in *Third International Conference on Informatics in Control, Automation and Robotics (ICINCO), the Second International Workshop on Multi-Agent Robotic Systems (MARS)* (2006), pp. 13–21
9. Y. Altshuler, V. Yanovsky, I. Wagner, A. Bruckstein, Swarm intelligencesearchers, cleaners and hunters. *Swarm Intelligent Systems* (2006), pp. 93–132
10. Y. Altshuler, I.A. Wagner, A.M. Bruckstein, On swarm optimality in dynamic and symmetric environments **7**, 11 (2008)
11. Y. Altshuler, V. Yanovsky, A.M. Bruckstein, I.A. Wagner, Efficient cooperative search of smart targets using uav swarms. ROBOTICA **26**, 551–557 (2008)
12. Y. Altshuler, I.A. Wagner, A.M. Bruckstein, Collaborative exploration in grid domains, in *Sixth International Conference on Informatics in Control, Automation and Robotics (ICINCO)* (2009)
13. Y. Altshuler, S. Dolev, Y. Elovici, N. Aharony, Ttled random walks for collaborative monitoring, in *NetSciCom 2010 (Second International Workshop on Network Science for Communication Networks), San Diego, CA, USA*, vol. 3 (2010)

14. Y. Altshuler, I.A. Wagner, V. Yanovski, A.M. Bruckstein, Multi-agent cooperative cleaning of expanding domains. Int. J. Robot. Res. **30**, 1037–1071 (2010)
15. Y. Altshuler, N. Aharony, M. Fire, Y. Elovici, A Pentland, Incremental learning with accuracy prediction of social and individual properties from mobile-phone data, in *CoRR* (2011)
16. Y. Altshuler, N. Aharony, A. Pentland, Y. Elovici, M. Cebrian, Stealing reality: when criminals become data scientists (or vice versa). Intell. Syst. IEEE **26**(6), 22–30 (2011)
17. Y. Altshuler, M. Fire, N. Aharony, Y. Elovici, A Pentland, How many makes a crowd? on the correlation between groups' size and the accuracy of modeling, in *International Conference on Social Computing, Behavioral-Cultural Modeling and Prediction* (Springer, 2012), pp. 43–52
18. Y. Altshuler, M. Fire, N. Aharony, Z. Volkovich, Y. Elovici, A. Sandy Pentland, Trade-offs in social and behavioral modeling in mobile networks, in *Social Computing, Behavioral-Cultural Modeling and Prediction* (Springer, 2013), pp. 412–423
19. Y. Altshuler, E. Shmueli, G. Zyskind, O. Lederman, N. Oliver, A. Pentland, Campaign optimization through behavioral modeling and mobile network analysis. IEEE Trans. Comput. Soc. Syst. **1**(2), 121–134 (2014)
20. Y. Altshuler, A. Pentland, S. Bekhor, Y. Shiftan, A. Bruckstein, Optimal dynamic coverage infrastructure for large-scale fleets of reconnaissance uavs (2016), arXiv:1611.05735
21. Y. Altshuler, R. Puzis, Y. Elovici, S. Bekhor, A. Sandy Pentland, On the rationality and optimality of transportation networks defense: a network centrality approach. *Securing Transportation Systems* (2015), pp. 35–63
22. J.M. Anthonisse, The rush in a directed graph. Technical Report BN 9/71, Stichting Mathematisch Centrum, Amsterdam (1971)
23. R.C. Arkin, Integrating behavioral, perceptual, and world knowledge in reactive navigation. Robot. Auton. Syst. **6**, 105–122 (1990)
24. D. Ball et al., *Burma's Military Secrets: Signals Intelligence (SIGINT) from 1941 to Cyber Warfare* (White Lotus Press, 1998)
25. M. Barthélemy, Betweenness centrality in large complex networks. Eur. Phys. J. B - Condens. Matter **38**(2), 163–168 (2004)
26. R. Bejar, B. Krishnamachari, C. Gomes, B. Selman, Distributed constraint satisfaction in a wireless sensor tracking system, in *Proceedings of the IJCAI-01 Workshop on Distributed Constraint Reasoning* (2001)
27. S. Bekhor, Y. Cohen, C. Solomon, Evaluating long-distance travel patterns in israel by tracking cellular phone positions. *J. Adv. Transp.* pp. n/a–n/a (2011)
28. M.G. Bell, S. Grosso, A. Ehlert, The optimisation of traffic count locations in road networks. Transp. Res. Part B **40**(6), 460–479 (2006)
29. D.J. Berndt, A.R. Hevner, J. Studnicki, Bioterrorism surveillance with real-time data warehousing, in *Proceedings of the 1st NSF/NIJ conference on Intelligence and security informatics, ISI'03* (Springer, Berlin, 2003), pp. 322–335
30. P. Bork, L.J. Jensen, C. von Mering, A.K. Ramani, I. Lee, E.M. Marcotte, Protein interaction networks from yeast to human. Curr. Opin. Struct. Biol. **14**(3), 292–299 (2004)
31. U. Brandes, On variants of shortest-path betweenness centrality and their generic computation. Soc. Netw. **30**(2), 136–145 (2008)
32. Z. Butler, A. Rizzi, R. Hollis, Distributed coverage of rectilinear environments, in *Proceedings of the Workshop on the Algorithmic Foundations of Robotics* (2001)
33. D. Byman, Why drones work. Foreign Aff. **92**(4), 32–43 (2013)
34. H. Chen, F.-Y. Wang, D. Zeng, Intelligence and security informatics for homeland security: information, communication, and transportation. IEEE Trans. Intell. Transp. Syst. **5**(4), 329–341 (2004)
35. R. Chesney, Military-intelligence convergence and the law of the title 10/title 50 debate. J. Natl. Secur. Law Policy **5**, 539 (2012)
36. D. Chevallier, S. Payandeh, On kinematic geometry of multi-agent manipulating system based on the contact force information, in *The Sixth International Conference on Intelligent Autonomous Systems (IAS-6)* (2000), pp. 188–195

37. G. Confessore, P. Reverberi, L. Bianco, A network based model for traffic sensor location with implications on o/d matrix estimates. Transp. Sci. **35**(1), 50–60 (2001)
38. L. Damianos, J. Ponte, S. Wohlever, F. Reeder, D. Day, G. Wilson, L. Hirschman, Mitap for bio-security: a case study. AI Mag. **23**(4), 13–29 (2002)
39. S. Dolev, Y. Elovici, R. Puzis, P. Zilberman, Incremental deployment of network monitors based on group betweenness centrality. Inf. Proc. Lett. **109**, 1172–1176 (2009)
40. A. d'Onofrio, A general framework for modeling tumor-immune system competition and immunotherapy: mathematical analysis and biomedical inferences. Physica D **208**, 220–235 (2005)
41. G. Dudek, M. Jenkin, E. Milios, D. Wilkes, Robotic exploration as graph construction. IEEE Trans. Robot. Autom. **7**, 859–865 (1991)
42. G.M. Erickson, P.J. Currie, B.D. Inouye, A.A. Winn, Tyrannosaur life tables: an example of nonavian dinosaur population biology. Science **313**(5784), 213–217 (2006)
43. M.G. Everett, S.P. Borgatti, The centrality of groups and classes. Math. Soc. **23**(3), 181–201 (1999)
44. M. Faloutsos, P. Faloutsos, C. Faloutsos, On power-law relationships of the internet topology. SIGCOMM Comput. Comm. Rev. **29**(4), 251–262 (1999)
45. A. Felner, Y. Shoshani, Y. Altshuler, A.M. Bruckstein, Multi-agent physical a* with large pheromones. J. Auton. Agents Multi-Agent Syst. **12**(1), 3–34 (2006)
46. L.C. Freeman, A set of measures of centrality based on betweenness. Sociometry **40**(1), 35–41 (1977)
47. B.P. Gerkey, M.J. Mataric, Sold! market methods for multi-robot control, in *IEEE Transactions on Robotics and Automation, Special Issue on Multi-robot Systems* (2002)
48. B. Gompertz, On the nature of the function expressive of the law of human mortality, and on a new mode of determining the value of life contingencies. Philos. Trans. R. Soc. Lond. **115**, 513–583 (1825)
49. T. Haynes, S. Sen, Adaptation and Learning in Multi-Agent Systems. *Evolving Behavioral Strategies in Predators and Prey*, vol. 1042, Lecture Notes in Computer Science (Springer, Berlin, 1986), pp. 113–126
50. I. Henderson, Civilian intelligence agencies and the use of armed drones, in *Yearbook of International Humanitarian Law-2010* (Springer, 2011), pp. 133–173
51. S. Hettiarachchi, W. Spears, Moving swarm formations through obstacle fields, in *International Conference on Artificial Intelligence* (2005)
52. P. Holme, Congestion and centrality in traffic flow on complex networks. Adv. Complex Syst. **6**(2), 163–176 (2003)
53. P.S. Kanaroglou, M. Jerrett, J. Morrison, B. Beckerman, M. Altaf Arain, N.L. Gilbert, J.R. Brook, Establishing an air pollution monitoring network for intra-urban population exposure assessment: a location-allocation approach. Atmos. Environ. **39**(13), 2399–2409 (2005). <ce:title>12th International Symposium, Transport and Air Pollution</ce:title> <xocs:fullname>12th International Symposium, Transport and Air Pollution</xocs:fullname>
54. W. Kerr, D. Spears, Robotic simulation of gases for a surveillance task, in *Intelligent Robots and Systems (IROS 2005)* (2005), pp. 2905–2910
55. S. Kirkpatrick, J.J. Schneider, How smart does an agent need to be? Int. J. Mod. Phys. C **16**, 139–155 (2005)
56. B.O. Koopman, The theory of search ii, target detection. Oper. Res. **4**(5), 503–531 (1956)
57. B. Koopman, *Search and Screening: General Principles with Historical Applications* (Pergamon Press, Oxford, 1980)
58. R.E. Korf, W. Zhang, Performance of linear-space search algorithms. Artif. Intell. **79**(2), 241–292 (1995)
59. J. Lerner, Role assignments. *Network analysis: Methodological Foundations*, LNCS 3418 (Springer, 2005)
60. H. Lo, W. Lam, Accuracy of o-d estimates from traffic counts. Traffic Eng. Control **31**, 358–367 (1990)

61. F. Lorrain, H.C. White, Structural equivalence of individuals in social networks. J. Math. Sociol. **1**(1), 49–80 (1971)
62. Lt. Kendra, L.B. Cook, The silent force multiplier: the history and role of uavs in warfare, in *Aerospace Conference, 2007 IEEE* (IEEE, 2007), pp. 1–7
63. H.S. Mahmassani, S.M. Eisenman, X. Fei, Sensor coverage and location for real-time traffic prediction in large-scale networks. Transp. Res. Rec. **2039**(1), 1–15 (2007)
64. M.J. Mataric, Designing emergent behaviors: From local interactions to collective intelligence, in *Proceedings of the Second International Conference on Simulation of Adaptive Behavior*, ed. by J. Meyer, H. Roitblat, S. Wilson (MIT Press, 1992), pp. 432–441
65. P.M. Morse, G.E. Kimball, *Methods of Operations Research* (MIT Press, Wiley, New York, 1951)
66. A.T. Murray, K. Kim, J.W. Davis, R. Machiraju, R. Parent, Coverage optimization to support security monitoring. Comput. Environ. Urban Syst. **31**(2), 133–147 (2007)
67. Y. Ouyang, X. Li, Reliable sensor deployment for network traffic surveillance. Transp. Res. Part B **45**, 218–231 (2011)
68. K. Passino, M. Polycarpou, D. Jacques, M. Pachter, Y. Liu, Y. Yang, M. Flint, M. Baum, *Cooperative Control for Autonomous Air Vehicles, chapter Cooperative Control and Optimization* (Kluwer Academic, Boston, 2002)
69. S. Peeta, C.-H. Chu, S.-R. Hu, Identification of vehicle sensor locations for link-based network traffic applications. Transp. Res. Part B **43**(8–9), 873–894 (2009)
70. J.T.K. Ping, A. Eng Ling, T. Jun Quan, C. Yea Dat, Generic unmanned aerial vehicle (uav) for civilian application-a feasibility assessment and market survey on civilian application for aerial imaging, in *2012 IEEE Conference on Sustainable Utilization and Development in Engineering and Technology (STUDENT)* (IEEE, 2012), pp. 289–294
71. R. Puzis, Y. Altshuler, Y. Elovici, S. Bekhor, Y. Shiftan, A.S. Pentland, Augmented betweenness centrality for environmentally-aware traffic monitoring in transportation networks (2013)
72. R. Puzis, M. D. Klippel, Y. Elovici, S. Dolev, Optimization of nids placement for protection of intercommunicating critical infrastructures, in *EuroISI* (2007)
73. R. Puzis, Y. Elovici, S. Dolev, Fast algorithm for successive computation of group betweenness centrality. Phys. Rev. E **76**(5), 056709 (2007)
74. R. Puzis, Y. Elovici, S. Dolev, Finding the most prominent group in complex networks. AI Comm. **20**, 287–296 (2007)
75. R. Puzis, Y. Altshuler, Y. Elovici, S. Bekhor, Y. Shiftan, A. Pentland, Augmented betweenness centrality for environmentally-aware traffic monitoring in transportation networks. J. Intell. Transp. Syst. **17**, 91–105 (2013)
76. G. Rabideau, T. Estlin, T. Chien, A. Barrett, A comparison of coordinated planning methods for cooperating rovers, in *Proceedings of the American Institute of Aeronautics and Astronautics (AIAA) Space Technology Conference* (1999)
77. E. Regev, Y. Altshuler, A.M. Bruckstein, The cooperative cleaners problem in stochastic dynamic environments (2012), arXiv:1201.6322
78. I. Rekleitis, V. Lee-Shuey, A. Peng Newz, H. Choset, Limited communication, multi-robot team based coverage, in *IEEE International Conference on Robotics and Automation* (2004)
79. P. Rouvinen, Diffusion of digital mobile telephony: are developing countries different? Telecommun. Policy **30**(1), 46–63 (2006)
80. N. Schörnig, Unmanned warfare: towards a neo-interventionist era? in *The Armed Forces: Towards a Post-Interventionist Era?* (Springer, 2013), pp. 221–235
81. J. Scott, *Social Network Analysis: A Handbook* (Sage Publications, London, 2000)
82. B. Shucker, J.K. Bennett, Target tracking with distributed robotic macrosensors, in *Military Communications Conference (MILCOM 2005)*, vol. 4, pp. 2617–2623 (2005)
83. C. Solomon, L. Kheifits, Y.J. Gur, S. Bekhor, Intercity person trip tables for nationwide transportation planning in israel obtained from massive cell phone data. Transp. Res. Rec. J Transp. Res. Board **2121**, 145–151 (2009)
84. L. Steels, Cooperation between distributed agents through self-organization, in *Decentralized A.I - Proc. first European Workshop on Modeling Autonomous Agents in Multi-Agents world*, ed. by Y. DeMazeau, J.P. Muller (Elsevier, 1990), pp. 175–196

85. R. Stern, R. Puzis, A. Felner, Potential search: a bounded-cost search algorithm, in *AAAI 21st International Conference on Automated Planning and Scheduling (ICAPS)* (2011)
86. S.H. Strogatz, Exploring complex networks. Nature **410**, 268–276 (2001)
87. S.M. Thayer, M.B. Dias, B.L. Digney, A. Stentz, B. Nabbe, M. Hebert, Distributed robotic mapping of extreme environments, in *Proceedings of SPIE, volume 4195 of Mobile Robots XV and Telemanipulator and Telepresence Technologies VII* (2000)
88. A. Thorndike, Summary of antisubmarine warfare operations in world war ii. Summary report, NDRC Summary Report (1946)
89. P. Vincent, I. Rubin, A framework and analysis for cooperative search using uav swarms, in *ACM Symposium on applied computing* (2004)
90. I.A. Wagner, A.M. Bruckstein, From ants to a(ge)nts: a special issue on ant–robotics. Ann. Math. Artif. Intell. Spec. Issue Ant Robot. **31**(1–4), 1–6 (2001)
91. I.A. Wagner, Y. Altshuler, V. Yanovski, A.M. Bruckstein, Cooperative cleaners: a study in ant robotics. Int. J. Robot. Res. (IJRR) **27**(1), 127–151 (2008)
92. J.G. Wardrop, Some theoretical aspects of road traffic research. Proc. Inst. Civ. Eng. **1**, 325–378 (1952)
93. S. Wasserman, K. Faust, *Social Network Analysis: Methods and Applications* (Cambridge University Press, Cambridge, 1994)
94. M.P. Wellman, P.R. Wurman, Market-aware agents for a multiagent world. Robot. Auton. Syst. **24**, 115–125 (1998)
95. D.R. White, S.P. Borgatti, Betweenness centrality measures for directed graphs. Soc. Netw. **16**, 335–346 (1994)
96. C. Yang, L. Gan, H. Yang, Models and algorithms for the screen line-based traffic-counting location problems. Comput. Oper. Res. **33**(3), 836–858 (2006)
97. S.H. Yook, H. Jeong, A.-L. Barabasi, Modeling the internet's large-scale topology. Proc. Natl. Acad. Sci. **99**(21), 13382–13386 (2002)
98. J. Zhou, H. Yang, Optimal traffic counting locations for origin-destination matrix estimation. Transp. Res. Part B **32**(2), 109–126 (1998)
99. S. Zilberstein, Using anytime algorithms in intelligent systems. AI Mag. **17**(3), 73–83 (1996)

Printed in the United States
By Bookmasters